Thesaurus Scienta Lancastriae
Robert Williams & *Jack Aylward-Williams*

Thesaurus Scientia Lancastriae

Robert Williams
&
Jack Aylward-Williams

UNIPRESS CUMBRIA

First published in Great Britain 2006 by UniPress Cumbria and
informationasmaterial

Editor: Robert Williams

ISBN 1-869979-19-2

Published with the support of Cumbria Institute of the Arts

Designed by Joakim Karlsson
Art direction by Anna Danby
Printed by Hillprint Media, Durham

4

CONTENTS

CONTRIBUTORS

Jack Aylward-Williams

Jack Aylward-Williams, the son of Robert Williams and Gina Aylward, was born in Lancaster on Christmas Day 1998. A natural collector, Jack's current interests include insects and natural history, fire fighting equipment, Harry Potter and Dr. Who; he has an encyclopaedic knowledge of the flora and fauna, and the anthropology and archaeology of Williamson Park in Lancaster. His collaborations with his father include a series of ten chimney drawings in 1999, *Thesaurus Scienta Lancastriae*, and a forthcoming art and archaeology exhibition at Penrith Museum in Cumbria in 2006. Many more collaborations are anticipated.

Jack's observations are marked in red throughout the publication.

David Barrowclough

David Barrowclough is a pioneering archaeologist with an interest in the relationship between art and archaeology. Based in the Archaeology Department of Cambridge University, he is the current editor of the periodical *Archaeological Review from Cambridge*. His present research is to identify, describe and document all extant bronze and iron age artefacts in the North West of England. David was the archaeological consultant for the BBC on the recent *How Art Made the World* television series.

Mark Dion

Mark Dion is one of the world's most significant contemporary artists, an American with a global reputation, his field of interest deals with representations of nature and ranges in subject matter from the pioneering days of natural history and the history of the natural sciences, to an investigation into the meaning of animals, wildlife conservation, museology, and archaeology. Recent work in the UK includes *The Tasting Garden* at the Storey Institute at Lancaster as part of *Artranspennine98* (1998), the highly influential *Tate Thames Dig* (1999), *Theatrum Mundi: Armarium* for the Cambridge Biennale in 2001, the *Bureau of the Centre for the Study of Surrealism and its Legacy* for Manchester Museum/University of Manchester (2005), and the highly acclaimed *Microcosmographia* at The South London Gallery (2005). He received the prestigious Aldrich prize in 2003 the same year that he received an honorary doctorate from Hartford Art School, the University of Hartford in Connecticut. He lives in Pennsylvania, USA.

Simon Morris

Simon Morris is the founder of informationasmaterial, an artist, theorist, and teacher whose practice deals with the use of textual and conceptual material already extant in the world. His notable works include the two highly influential *Bibliomania* projects (1998-1999 and 2000-2001), a series of collaborations, events and performances with the psychoanalyst Howard Britton such as *Spinning* (2003) and *The Royal Road to the Unconcious* (2004) at the Freud Museum, and recently the publishing of an ambitious re-presentation of Freud's book *The Interpretation of Dreams* (2005). Simon teaches Art History, Theory & Criticism at York College.

John Rodwell

John Rodwell coordinated the research team that produced the first systematic classification of the vegetation of this country and he edited the five-volume *British Plant Communities*. He founded the Unit of Vegetation Science at Lancaster University and was made Professor of Plant Ecology there in 1997. He is now an independent consultant ecologist, working with such organisations as the Forestry Commission, Defra, English Nature, the National Trust and Groundwork to understand existing and potential landscapes and their significance to the people who live in them. He himself lives in Lancaster within sight and sound of Williamson's Park.

Peter Wade

Peter Wade is an historian of science currently working at Lancaster University in the Department of Continuing and Adult Education. His research into Lancaster's scientific heritage has made a significant contribution to an understanding of the early lives of pivotal nineteenth century scientists native to the town, and of popular, social and cultural attitudes to science during the period. As well as contributing to the publications from Lancaster City Museum, Peter organises occasional tours of locations in the city linked to figures such as Whewell, Frankland, Turner, and of course, Sir Richard Owen.

Robert Williams

Robert Williams is an artist and academic, having trained at Lancaster University and at Leeds University where he was a Henry Moore scholar in Sculpture Studies in 1990. He has been leader of the Fine Art programme at Cumbria Institute of the Arts since 1998. Robert's interdisciplinary practice encompasses sculpture, installation, performance, film-making, and writing, and incorporates an interest into epistemology and systems of knowledge from the hermetic to the scientific. His recent practice includes collaborative projects with Mark Dion in the UK and USA including *The Tasting Garden* at Lancaster (1998), *The Tate Thames Dig* (1999), *Theatrum Mundi: Armarium* (2001), Mark Dion: *Collaborations* at Hartford (2003), *Down the Garden Path: Artists Gardens After Modernism* at the Queens Museum of Art, New York (2005) and a series of prints with Dion and students from Fine Art at Cumbria Institute of the Arts which formed part of the Underground Art project associated with Dion's *Microcosmographia* one man show at the South London Gallery in October 2005.

PREFACE

Mark Dion

Thesaurus Scienta Lancastriae

Specimen: Tree – Wych Elm
Date: 23/10/04
Location: Williamson Park
Robert Williams & Jack Aylward-Williams

11/3/04
magpie

Thesaurus Scienta Lancastriae

Specimen: Bird's Feather – **Magpie** *(Pica Pica)*
Date: 29/8/04
Location: Williamson Park
Robert Williams & Jack Aylward-Williams

'The specimen you collect should be selected with care. From a dozen that seem likely, keep only a few. Hand-sized specimens are preferred. Some collectors look for small, perfect, thumbnail-sized specimens and study them with a low-power microscope. Be sure your specimen is fresh. Trim it to shape.' (Zim & Schaffer, 1957)

'Most people collect for the simple fun of it – for the fun of tramping and exploring; for the excitement of the rare find; for the challenge of 'working out' a perfect specimen. But in the course of doing all this, the layers of sedimentary rocks unfold like pages of a gigantic book, revealing the fascinating study of earth's long and extinct past. Events 50, 100 or 500 million years ago become real because the fossils you have found provide a clear connection with bygone ages.' (Rhodes, Zim & Schaffer, 1962)

Thesaurus Scienta Lancastriae is a complex work in which Jack Aylward-Williams, all of six years old, explains the world to his father, Robert Williams, who translates Jack's cosmology to us. Jack is the scientist, Robert his museum. The project's field of investigation encompasses one of the most profound questions in the history of ideas – how do we know what we know?

Sculptors, like children, are magnetically drawn to the culture of collecting and to museums – places where we obtain knowledge through contact with things. Visitors flock to museums, not to view a picture, or depiction of 'the thing', not to watch a video, or computer image of 'the thing', but rather to encounter 'the thing' itself. This object is, of course, a representation, framed and contextualised. However, it is a very particular variety of representation, one that shares our scale, space, and moment in time. This thing must bear vast metaphoric weight. Astoundingly intricate narratives or concepts must be embodied in an artefact or specimen, which explains the interest of sculptors like Williams in the gathering of material culture. For young Jack, the shards, feathers and stones that he collects are clues to the *Grand Mystery*, he confidently elucidates to his father, much in the way Sherlock Holmes admonishes Dr Watson for missing vital facts.

To those not predisposed to the phenomenon of collecting, it may be unsettling to witness the sheer joy a novel acquisition can inspire in a collector, the crushing anxiety a lost or broken object can provoke. For most, the external world need not be manifested through things to be more deeply comprehended. However, for those of us devoted to artefacts, archives, specimens, ephemera and display technologies, the world is expressed, expanded and, paradoxically, coalesced by these objects; they are the keys to understanding.

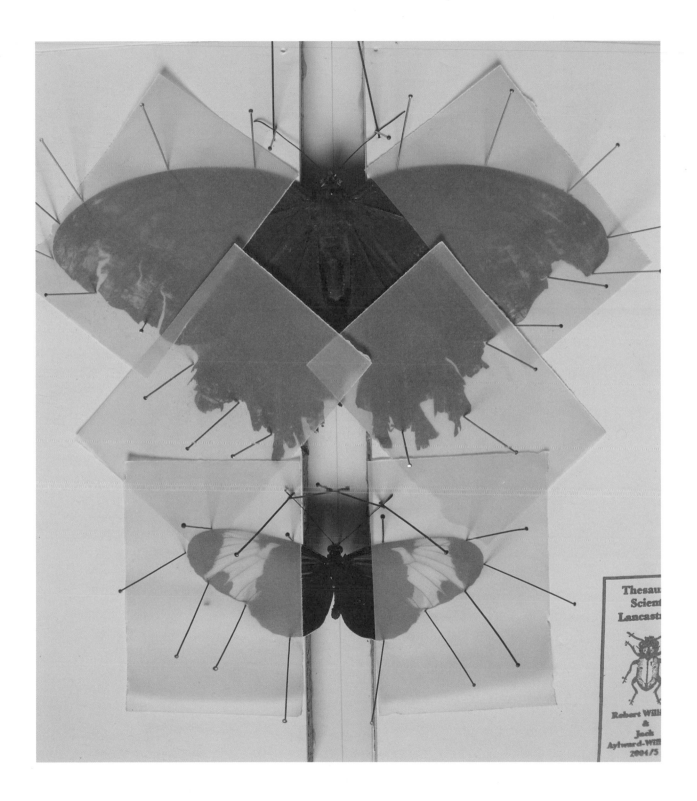

BEETLE ELYTRA
A beetle has very shiny wing cases.

WATER BEETLES
Water beetles have wings like cicadas and so do cockroaches. Cicadas have wings as big as cockroaches. Cicadas have three wings, two on the top and one the bottom. Cockroaches have four wings. All creepy crawlies that are as big as cockroaches, cicadas or water beetles have three or four wings – except wasps.

By selecting, amassing, storing and displaying, collectors like Jack construct microcosms – diminutive worlds of subjective value. The *Thesaurus Scienta Lancastriae* microcosm already exists in the 19th century vision expressed by Lancaster's Williamson Park. Yet how to capture it is the problem. There are potentially millions of things to gather – the leaves of trees, ants, plastic bottle caps, footprints, stones. How does Jack know what is worthy of collecting? What kind of external world can we aliens imagine based on Jack's taxonomies and remarkable field observations? Where does Jack's absolute certainty of his interpretations derive from?

These questions we could easily ask of the Manchester Museum, Victoria and Albert or British Museum. Jack is every bit a botanist, zoologist, geologist, anthropologist, archaeologist and meteorologist, and, like many scientists, he seems best able to understand the world by holding it close. He gains possession of his universe by securing jigsaw puzzle pieces, which eventually may constitute a comprehensive picture. Walter Benjamin clearly understood collectors like Jack and Robert when he wrote:

> 'Ownership is the most intimate relationship one can have to objects. Not that they come alive in him; it is he who lives in them.'
> (Benjamin, 1931)

Of course, Jack's field of investigation is the world, while Robert's field is Jack. I cannot imagine a more in-depth or sophisticated method of probing the formation of subjectivity than the physical manifestation of a world view. Robert Williams has produced a work motivated by deed curiosity, but manifested through love.

Mark Dion
October 2005

References

Benjamin W (1931) *Illuminations.* London: Fontana, 1973

Rhodes FHT, Zim HS, Schaffer PR (1962) *A Golden Guide to Fossils.* New York: Golden Press.

Zim HS, Schaffer PR (1957) *A Golden Guide to Rocks and Minerals.* New York: Golden Press.

INTRODUCTION AND ACKNOWLEDGEMENTS

Robert Williams

Dedication:
For my Father, Jack's Grandfather, Bob Williams: a hard act to follow, but I'm trying to be just like you.

A fascinating aspect of the City of Lancaster is the number of very influential nineteenth-century scholars and scientists native to the town, among whom are included Sir Richard Owen, William Whewell and Sir Edward Frankland. These figures, and others discussed in Peter Wade's contribution to this book, represent a particular world view; a way of observing, ordering and organising observed phenomena. This positivist-empirical, or scientific approach to an experience of the world – the ordering of relationships, causalities and taxonomies, and the creation of interpretations of observed phenomena, continues to inform our society and culture. Yet it is arguable that the basis for this organisation is based less upon that empirical method, and more upon idiosyncratic, and intuitive reasoning, and even poetic interpretation, as personalities like Richard Owen tried to reconcile their beliefs and background to the new data. Also part of a nineteenth century world view was the creation of public spaces dedicated to a consideration of cultural values such as education, spiritual well-being, edification, morality and good works, exemplified by local notaries such as James Williamson II, Lord Ashton, the man who built the Ashton Memorial in the park created by his father James Williamson. Within the Williamson Park are further examples of the fascination in and enthusiasm for scientific observations and collections which become linked to a high culture, in the form of edifying architecture and garden design culminating in Sir John Belcher's fabulous memorial to the Williamson family. The Greg Astronomical Observatory and weather station, and the Kew-like plant collections in The Palm House (now the Butterfly House) all contribute to this cultural dynamic.

Thesaurus Scienta Lancastriae as a project which encompasses not only the installation and its collections but also this book, seeks to explore these social, cultural and critical issues by using Williamson's Park as a microcosm of the world onto which can be projected a collecting and organisational activity resulting in a work of art.

Fundamentally, the project is collaboration between myself and my son Jack. My role is to act as facilitator, curator and organiser of collections that he generates and interprets. The Williamson Park contains a wide range of collecting contexts all of which reflect these nineteenth century obsessions of science and culture: areas explored within the environments offered by the Park, as the title of the project suggests, include geology and palaeontology, botany, biology, zoology, physics, chemistry, meteorology, astronomy, arboriculture, anthropology, and many, many others. That the collecting

ANT
Unlike other ants, which have mandibles, black ants move things with their lips, they don't have mandibles.

phase of the project occurred during the bi-centenary of Sir Richard Owen's birth, was no accident, and his connection to the project is signalled by a mounted specimen of the Pearly Nautilus (*Nautilus Pompilius*), a species that he identified and described in 1832.

Since Jack was small, I have been fascinated by the way in which he was able to create sophisticated theories which have enabled him to successfully navigate his world. True to Carl Popper's classic *Hypethetico Deductive Method* (Popper 1972) for a philosophy of science, Jack would hold on to a theory until a new one came along which invalidated, or helped to modify it:

> 'Earwigs are rare – but they can swim like sandhoppers. They are like sandhoppers.' 20/7/04

> 'On earwigs, the pincers, where spinnerets usually are (on spiders), these are called 'spinners'.' 21/7/04

> 'They look like sandhoppers and they have not 'nippers' but swimmers on their end. They look like spinnerets on a spider. Earwigs can swim.' 16/8/04

> 'Boy earwigs have swimmers, like spinnerets. Girl earwigs have forceps.' 22/8/04 (Aylward-Williams 2004)

Then he got on with the job of being a boy. Nevertheless, like many scientists before him, Richard Owen is a prime example, Jack would also sometimes doggedly hold onto his former ideas despite the advantages provided by the modified circumstances (Cadbury 2000).

At the beginning of my thoughts about the project, I faced issues of my own, the need to deal with the incredible demands of a full-time teaching position and academic leadership of a programme of study, to maintain my research and art practice, and to be a father – all seemingly conflicting calls on my attention. It seemed reasonable to take the line of least resistance and to find a way to work with these demands.

I was no stranger to collaboration, my work with Mark Dion since 1998 (Barley 1999) had shown me just how challenging and satisfying this strategy for practice could be, and it was, therefore, a tiny step to modify the usual activities that Jack and I often enjoyed together collecting bits and pieces, making drawings, making experiments, having conversations about the natural world – to arrive at a practice to which we could both equally contribute in our different ways. My experience of Dion's model for practice has been that he invites his collaborators to work with him according to their own interests and strengths (Williams 1999), and to allow this dynamic to feed the work. Jack and I decided that this too would form our approach, he taking the role of collector, and me the archivist and curator. Our mutual interest in museums and in collecting underpinned the plans for the project – Jack would, as a baby and toddler, 'collect' streetlamps, electrical plugs, and memorably drainpipes (we have a book of images, redolent of Bernd and Hiller Becher's work, that needs to be published); becoming extraordinarily enthusiastic toward each of these benign and mundane objects that we encountered, so much so that a five minute journey could last over an hour as we stopped to admire the object of his exclusive interest as (and when) we encountered it.

Jack's interests, as he grew, became portable – sticks, polished stones, pinecones, for example; and so his collecting practice matured into not only gathering, but in the making of collections. That my wife, Jack's mother, Gina Aylward, and I shared this collecting passion, and supported it, perhaps obscured the clear signals of Jack's Autistic Spectrum Disorder (ASD). Indeed, my plans for the project – recognising Lancaster's nineteenth century scientific heritage from my reading of Peter Wade's astonishing notes from Lancaster City Museum, revised and re-presented here, and Andrew White's fascinating *History of Lancaster* (White 2002), were already in place by the time Jack was three years old. We had visited Williamson Park (Ashworth. 2002) from the very early days, and had been intrigued by the ruins as we called them – not realising that these were the foundations of the lost Greg Astronomical Observatory (Bone 1892). Jack and I would project our fantasies onto these inherently romantic ruins, making of them evidence for a lost civilization, an archaeological site, a Roman fort, the house of boy from the past, and more. Again, Peter Wade's study of the lost observatory helped to make these ruins much more interesting than we ever imagined. David Barrowclough wonderfully frames the archaeological imagination at work in projecting our interpretations and emphases of interest on to the landscape of the park, the ruins, traces, objects and creatures that we encounter. Barrowclough's text helps to underpin the importance of creating the interpretive spaces made available in not imposing authoritative labels or captions, and instead allowing for an opportunity to speculate, to invest and open up the kind of interpretive possibilities used to such advantage by Jack in his experience and navigation of the world.

STAG BEETLES

If you (Robert) or me were stag beetles, we'd have big antlers that don't hurt very much. But if Mummy was a stag beetle, she'd have mandibles that really hurt 'cos a male stag beetle couldn't bite through a pencil, but a female could.

SQUIRRELS AT WAR
Grey squirrels are not very nice to the
red squirrels: They don't like each other,
they throw acorns at each other. Because
every time one of them throws an acorn,
the other one throws it back until one of
them eats it up.

The plans for the project lapsed at the point of Jack's Statement for Asperger's and the change in the focus of our lives that this caused, and yet from this situation came the dawning realisation that whilst difficult for a parent and school to deal with, the consequences of his personality from Jack's point of view, were all positive – it is the rest of the world which is illogical, badly organised and confusing. Once Gina and I had begun to deal with the situation, there seemed even more reason to continue with the project, given what we had learned of ASD and the suggestion that many scientists exhibited evidence of this personality, Newton, Darwin, Einstein, and Richard Owen – indeed it seemed providential that Owen's bicentenary approached, providing Jack and me with a perfect vehicle to contextualise and frame the project. The project as an activity, began to raise issues of maleness – certainly in our relationship towards each other, and in encompassing the trend for boys to exhibit ASD in their behaviour – Simon Morris, John Rodwell and Mark Dion in their moving and insightful texts all imply this maleness in collecting and taxonomic list making as significant, and moreover, as a positive. These are traits that I can recognise in myself too. This is the simple explanation for the existence of the shed, as a particular space reserved for expressions of maleness as a spectrum of behaviours and practices as clichéd irony – every man needs his shed, so it is said; but also as a practicality (we needed somewhere to put all this stuff), and that this was always planned to be shown initially at the Ashton Memorial (Williams 1998), a humble venue in such an elaborate setting seemed somehow appropriate. However, the project, as it developed, whilst acknowledging these strata as elements in the process to bring the collections together, is not at all about ASD, or indeed gender politics, in how the work functions on either an emotional, contextual, or especially on a cultural level. Consequently, I am indebted to Simon Morris for his astute analysis of the project as art in this respect,

given my long time interests in epistemology, and in the construction of knowledge and belief systems, and indeed how these might be dealt with in an interdisciplinary approach to art practice and the production of an artwork that crosses boundaries in physical, conceptual and critical terms. In this book then, we find a document for the project, one that is photographic, textual, and interpretive, and which is meant to co-exist with the reality of the collections: the specimens, library, field equipment, the photography and data archive, and the audio and video archive in the installation itself.

Acknowledgements

Along the way, Jack and I have received the generous support and interest of an enormous number of people, and we should like to thank all of them for helping to make the project and the book possible. The greatest support, and invisible third member of the collaboration was Gina Aylward, without whom neither Jack nor I could exist, and who has worked tirelessly to enable us both to indulge ourselves. We are grateful to all our authors, David Barrowclough, Mark Dion, Simon Morris, John Rodwell and Peter Wade, for providing us with such astute readings of the project, and whose efforts have helped to move the project on during the process of becoming, in offering an interpretive and critical context to explore the work.

Thanks are due to Elaine Charlton and her wonderful staff at Williamson Park, all of whom have been enthusiastic and astonishingly supportive in facilitating the project and allowing us the run of the park, and the Ashton Memorial, the venue for exhibition; thanks especially to Graham Marshall for his enthusiasm and for the provision of exotic corpses from the MiniBeast House, many of whom we knew and loved in life.

Particular thanks are due to Michael Coombs, for being a brilliant and tireless assistant, and to his father – Peter Coombs for his advice on the building of the complex chassis, which has worked beautifully (and safely) as befitting a Formula One shed.

Our gratitude too, for the support and advice from Heather Dowler at Lancaster City Museum, and staff from Lancaster City Reference Library, and at the Photographic Unit at Lancaster University, for the historical material that they have generously provided.

Extra special thanks are due to Jack's school, Christchurch, and especially to headteacher Sandra Hall and Jack's TS3, Ian Wood, whose patient support for Jack and his interests have undoubtedly helped his confidence as well as his academic development. Thanks to Tony Bland for his invaluable support of Jack and for his advice in dealing with ASD.

We appreciate very much the support and encouragement from our sponsors during the collecting phase of the project – Reg Stoddon of Robertson's Photographic for his invaluable advice, Henry Smolarek of Lancaster PC for the website and generous support, Andrew Ireland of Lamberts (Lancaster) Optical Instruments, and Tim Wilton-Morgan of Timstar Laboratory Supplies.

THE PROJECT
Our plan is to look for flowers and insects (signs of Spring), we need to look for flowers.

FOXGLOVES & PITCHER PLANTS
Foxgloves eat flies as well as pitcher plants [do]. Because they look just like pitcher plants.

A debt of gratitude is owed for support from my colleagues in Fine Art and Graphic Design at Cumbria Institute of the Arts, Mark Wilson, and Glen Robinson, whose advice for the book design has been invaluable; and to Kevin Phillips, Nathalie de Briey, Dr. Julie Livsey and Martin Fowler for their enthusiasm and encouragement when things got tough. Thank you for positive support and feedback from Ian Farren, Director for the School of art and design at Cumbria Institute of the Arts and also for his first reading of the texts.

Finally, thanks are due to Karen Bassett of Research and Creative Enterprise Services and Charles Mitchell Chair of the Research and Scholarly Activity Advisory Group at Cumbria Institute of the Arts, and to Terry Kirton and the design team at UniPress, Joakim Karlsson, Anna Danby, Steve Minto, and Malcolm Farrar, for the book itself.

References

Ashworth S (2002) *Nature's Daintiest Bounty: Williamson Park and The Ashton Memorial.* Lancaster City Museum Pamphlet PLU113.

Aylward-Williams J (2004) Extracts from Audio Archive, *Thesaurus Scienta Lancastriae.*

Barley N (ed) (1999) *Leaving Tracks: artranspennine98* – An international contemporary art exhibition recorded. August Media artranspennine98

Bone J (1892) *The Greg Observatory: What may be its aims and objects.* Lancaster Philosophical Society, October 25 1892.

Cadbury D (2000) *The Dinosaur Hunters.* London: Fourth Estate/ Harper Collins.

Popper KR (1972) *Objective Knowledge: An Evolutionary Approach.* Oxford: OUP.

White A (1993) *A History of Lancaster: 1193–1993.* Keele: Ryburn/ Keele University Press.

Williams R (1998) *Herbert Hampton and The New Sculpture in Lancaster – The Victoria Monument and the Ashton Memorial.* In Roberts E (ed) *Regional Bulletin – Centre for North-West Regional Studies, New Series No 12,* Summer 1998. Lancaster: Lancaster University.

Williams R (1999) *Disjecta Reliquiae: The Tate Thames Dig.* In Coles A, Dion M (eds). *Mark Dion: Archaeology.* London: Black Dog.

Thesaurus S.

Weather S

Date	Time	Barometer		Thermo
D/M/Y		Mb	Mm	C
17/10/04	10:00am	980	735	8
cloudy, but dry – slight				bre
18/10/04	9=20	986	740	6°
Clear blue sky, some cloud coming from				
19/10/04	9=30	981	736	7
full cloud cover, low clouds some very				
20/10/04	9=45	973	730	7
full cloud cover, drizzly rain, slight b				
21/10/04	10am	972	728	7
Strong galeforce winds with rain, he				
22/10/04	10am	980	735	8
Heavy cloud cover moderate b				
23/10/04	10=30	975	732	8
Heavy cloud Slight breeze				
24/10/04	09·40	972	729	11
Grey & unsettled – light winds				
25/10/04	11·00pm	986	738	8·
Unsettled – light to moderate wind				

Shee

a Lancastriae

n Readings

	F	Rain Guage		Wind Dir	
		mm	inches	N/S/E/W	
	46	1.5	0.06	NE	1
e.					
	43	0	0	N	2
dry air. very slight breeze				
	44	5	0.2	S	3
e. Strong breeze (Southerly)					
	44	15	0.6	NW	4
ze					
	44	10	0.4	S	5
cloud cover					
	46	11.5	0.45	S	6
e. steady rainfall					
	46	38	1.5	S	7
amp air					
	52	3.5	0.75	S	8
	58	5	0.2	E	9
some rain					

nber: 7

Chapter 1. FROM THE POLITICAL TO THE PERSONAL

Simon Morris

Nineteenth-century scientists and scholars had a particular way of relating to the world: observing, selecting, cataloguing, and interpreting the *bric-a-brac* of life. Collections were formed on empirical research principles (observation and experimentation), which worked to establish new epistemologies. Regardless of the success or failure of these theoretical constructions, the collections tell us something else. The choice of objects and the method of their display create an image of the collector. *Thesaurus Scienta Lancastriae* is no different: it gives us an image of Robert Williams and his son Jack.

The working methods of Lancastrian scientists in the 19th century provide the framework for this project. Robert and Jack have set out to mimic these 19th-century activities in their year-long investigation of their local park. Together they have generated their own collection of diverse material – Robert Williams acting as facilitator, curator and organiser of the collection, which Jack generates and interprets.

I was introduced to the artist Robert Williams by the American artist, Mark Dion. The two artists, Williams and Dion, have worked together on numerous occasions and share some of the same methods, but I am going to suggest that the intention of their respective bodies of work is different. Having established that difference, I am going to examine the work through the framing that surrounds it.

The artist as researcher or the artist performing quasi-scientific procedures has a history, which, like a entomological specimen, can be pinned down. To trace this type of activity, you could travel as far back as Walter Benjamin's essay *The Author as Producer* (1934) or perhaps, more recently, Joseph Kosuth's *The Artist as Anthropologist* (1975) To consider work in the expanded field, you could consult the writings of Robert Smithson or look at the fictional museums of Marcel Broodthaers.

To reflect on the critical position of these practices, you could do no better than read James Meyer's *What Happened to the Institutional Critique?* (Meyer, 1993). In this essay Meyer cites further texts that were influential on a group of artists, some of whom came to masquerade as scientists/researchers: Craig Owen's *The Allegorical Impulse*, Rosalind Krauss's *Notes on the Index*, Douglas Crimp's *Pictures* and Benjamin Buchloh's *Allegorical Procedures: Appropriation and Montage in Contemporary Art*. Meyer credits these texts with 'provid[ing] the initial intellectual terrain for many of these artists'. Subsequent texts, such as Hal Foster's *The Artist as Ethnographer*, have further explored the idea of the artist as researcher.

You could also visit the seat of learning where these teachers practised, their texts were disseminated and their students put the critical strategies they had assimilated to good use, namely, the Whitney Independent Study Programme in New York. Joseph Kosuth, Rosalind Krauss, Benjamin Buchloh and Craig Owens have all taught

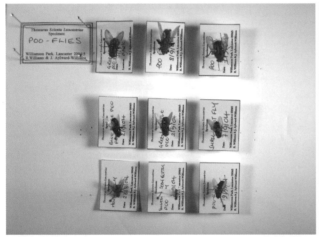

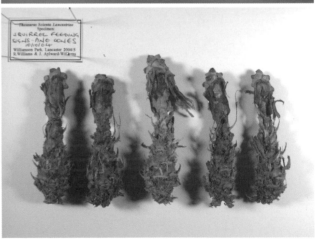

on this programme. (I've often thought you could have some great exhibitions of teachers and their students: Bernd and Hilla Becher and young German photographers, John Baldessari and West Coast conceptualism, Michael Craig-Martin and the young British artists, Craig Owens and the Whitney gang).

These are the texts and the sites in which you can locate the ideological foundations of this particular model of working. The teachers on the Whitney programme gave their students a thorough grounding in critical strategies of intervention, but it was what the students choose to map their 'conceptual tool box' (Coles, 1999:47) onto that marks out their individual practices as innovative and distinct from one another:

- Greg Bordowitz onto the politics of the AIDS crisis
- Mark Dion onto the systems of representation in natural history
- Andrea Fraser onto the pedagogic function of the museum
- Renée Green onto identity politics.

Mark Dion describes the point in his career when he decided to map two separate areas of his life together in his interview with Alex Coles in *The Optic of Walter Benjamin*:

> 'This transformation in my practice was also a result of my desire to bring together my own personal passions, which comes from a place too complex even for me to necessarily get entire control over it. On the one hand, I was living the life of someone who has a great passion for interacting with nature, and a great passion for its forms of representation, whether that be the Natural History Museum, television, or the apparatus of collecting, and on the other, I was living the life of an artist who was almost disinterestedly pursuing the methodology of institutional critique, something I didn't necessarily give a damn about, one way or the other. The important thing was I was able to bring together something I did give a damn about, with the same conceptual toolbox I had developed to take apart these other things. For me, this makes the strongest art in some way: where there is a level of conviction that comes from another place, not internal to art, that cannot be taught as an academic procedure.'

Dion goes on to credit the writing of Stephen Jay Gould that allowed him to see how critical methodologies could be mapped on to his interest in issues of extinction and conservation. (Coles, 1999:47)

As Hal Foster, Rosalind Krauss, Yve-Alain Bois and Benjamin D Buchloh state in their recent book, *Art Since 1900*:

> 'Each of these artists complicates the ethnographic approach with other models: Fraser is interested in the sociology of art pioneered by the French sociologist Pierre Bourdieu; Green in the postcolonial discourse of critics like Homi Bhabha; Dion in the study of disciplines developed by Foucault; and so on (Foster *et al*, 2004:629).

This type of work blurred the usual distinction between artist and curator. These artists also eradicated boundaries between their art-making and other activities in what became known as 'an interdisciplinary methodology'. In dissolving borders,

BUTTERFLIES
Butterflies have two feelers and two eyes. A snail or a slug has four eyes. The moth doesn't have any eyes. A butterfly is like a snail or a slug because its eyes are at the end of feelers.

FISHING NETS
Fishing nets are only to look at fish – they are not [used to catch fish] to eat them. It is fishing rods to eat them.

SPIDERS

Spiders can't spin webs – they can only spin one of the pieces of the web, a fine one; and when you are on a balcony and you find a spider, all you have to do is to wait for it to spin a piece of web. Take it off when it has spun it, and then the spider spins its way all the way to the bottom.

Tarantulas (from the Minibeast House) spin webs just like other spiders, but they don't catch their prey in the webs. They pounce out of their webs to catch their prey.

the students went further than their teachers: where the teachers saw a distinction between their art practice and their teaching, the students viewed all their activities as indissociable and continuous.

One of the artists to emerge from this particular paradigm, taught by both Owens and Kosuth, was Mark Dion. Meyer (1993:15,25) refers to Owens as a 'singular influence' and Kosuth as 'another important influence'.

In his work, Mark Dion has travelled the world and, by invitation, put the institutional systems and structures of representation in galleries and museums on display (according to what is, by his own admission, 'a Foucaultian model').

> 'Foucault's strategic acknowledgement of the unacknowledged frame of the university, his unmasking of its political imbrication, was soon adopted by artists who wished to unmask the interests at work in the institutional frames of the art world.' (Foster *et al*, 2004:548)

Dion's practice, like all practices, has a traceable theoretical foundation, but it is the dogged determination of his large-scale constructions on an international stage that has seen his work gain recognition as a serious and viable concern. I am reminded of the words of the American artist Kenneth Goldsmith:

> 'John Cage said something to the effect that anyone can do his work, but the fact is nobody else has done it. I take this to mean that the artist's real work is in setting the parameters and executing a given project. It's about the courage to actualise ideas that transform passing thoughts – often trivial – into art.' (Perloff, 2003)

It is one thing being introduced to particular ways of thinking, but it is what you do with that knowledge that matters – how you give your ideas form. And what forms – Dion has staged brilliant, large-scale, site-specific interventions across the world from the Venezuelan rainforests to the canals in Venice. His work has challenged existing display methodologies and representational systems within museological settings. Using his own language, he has successfully turned the museums inside out – putting the back rooms on exhibition and the displays into storage.

> 'The museum needs to be turned inside out – the back rooms put on exhibition and the displays put into storage.' Mark Dion in conversation with Miwon Kwon (Corrin *et al*, 1997:18)

While doing this on an international level, he has foregrounded the benefits of collaborative practices, a working model that allows him 'to take a vacation from himself'.

MOSSES LIVERWORTS AND

PAINTING

ARCHITECTURE

FRESHWATER FISH

GEOLOGY

GARDEN FLOWERS

TREES

MODERN ART

WILD ANIMALS

BIRDS

HEATH AND WOODLAND BIRDS

BIRDS OF PREY

Garden Birds

WHAT TO LOOK FOR IN WINTER

WHAT TO LOOK FOR IN AUTUMN

AL HISTORY

BRITISH WILD ANIMALS

THE STORY OF MEDICINE

MICHAEL FARADAY

GARDEN FLOWERS

THE SEASHORE and Seashore Life

The Story of ou... CKS AND MINE

STONE AGE MAN IN BRITAIN

OUR LAND IN THE MAKING

RVATION

THE WEATHER

ANIMALS, AND HOW THEY LIVE

Wild Life in Britain

POND LIFE

PREHISTORIC ANIMALS AND FOSS

AL HISTORY

Sea and estuary birds

Mark Dion on collaboration

'One great thing about collaboration is that it's like taking a vacation from yourself, if you're honest about it. I have a way of doing things and other artists have their way of doing things, and I learn a lot from that. Sometimes methods are very contradictory and it has to be their way or my way. It can be a struggle, things turn out differently. If I design a collaboration and it comes out exactly the way I thought, then it wasn't a productive collaboration. If it looks nothing like how I imagined it would look then it is really successful. The best test for me, personally, is how much the idea evolves with the influence of another person. My collaborators have always been strong personalities with definitive positions, and, so while it is always rewarding, it is not always easy. Some collaborations are also simply good excuses to travel and spend productive time with friends. We enjoy working together even if it is a challenge.' (Dion, 2003:3–4)

Several of these vacations have included his friend, the English artist Robert Williams. Mark Dion describes why he so much enjoys working with Robert Williams:

'Robert Williams has been a remarkable and delightful artist to collaborate with. We share a rather dark sensibility, a passion for the heyday of natural history, an interest in superstition and religion as symbolic visual systems. We share numerous other interests – such as a taste for Scotch whisky – but they are perhaps too many to detail.

Rob's dexterous mind overflows with an encyclopaedic knowledge of everything from obscure 16th century Latin incantations, to flint-knapping tool construction techniques, to kitsch horror film trivia. The various masses of arcane information are organised in his furtive brain in an utterly unhierarchical system, which contributes to his unconventional methodology as an artist.

While Rob is consistent in his commitment to the notion of art production as primarily an intellectual endeavour, he is also skilled as a maker. He easily masters the techniques he requires to make his work, regardless of whether it leads him to the bronze foundry, the computer editing suite, the carpenter's studio, or out into the field. He is the rarest of beasts – an artist as skilled with his hands as with his head.

Having known Rob for quite a few years now, I have had ample opportunity to enjoy his quick wit and dark humour. One of our chief motivations for working together is that we enjoy spending time together, swapping jokes and ghost stories. We share so many interests, fascinations formed in childhood, which oddly persisted, rather than dwindling, in adulthood. Working with Rob has been a blast, he is just a pleasure to be around. We work hard, but take great pleasure in the activity.

It is wonderful to collaborate with someone who shares your framework of reference, who gets the jokes no one else seems to fathom. It helps that we have read widely in the same territory, enjoyed many of the same films, and share an interest in many of the same artists.' (Dion, 2005)

One of my own teachers – whose influence resonates in me to this day – Dr Jane Rendell said:

'I discover parts of myself in my encounters with others.' (Rendell, 2002:53–54)

Both Williams and Dion have benefited from their encounters with each other. Williams made the technically accomplished bronze casts of fruit for Mark Dion's 'Tasting Garden' at the Storey Institute in Lancaster. You can see visible elements of Dion's practice in Williams' chosen methodology for *Thesaurus Scienta Lancastriae*. The white laboratory coats take us back to the *Thames Tate Dig* (in which Williams actively participated, both working at the field station and writing a text for the accompanying catalogue). The mobile shed can be linked to Dion's various huts, shacks and mobile field units.

Some of the vocabulary may be the same, but the intention and purpose of the work are different. Dion's work is art in praxis… the artist depicts society at the same time as he or she attempts to change it. The Williams and son's work is not so politically motivated or consciously subversive. Their work is about making a subjective space… a space they can take possession of.

Thesaurus Scienta Lancastriae

Specimen: Artificial Plants (various species)
Date: 16/8/04
Location: Williamson Park
Robert Williams & Jack Aylward-Williams

SCORPIONS

Scorpions are part sea creatures – part crabs, part bees, and part snails. Spiders can spin webs, and so can scorpions.

Surely it is no accident that the site Williams and his son have chosen for their exhaustive investigation is called Williamson (William(s)on) Park. This is their own space, which they have decided to investigate – did they have a choice? Perhaps their work is equally subversive in a different way in that it does not counter politics with more politics, or existing meaning with further meaning. They create their own subjective space outside of the political, which could be an example of what Michel Foucault refers to as non-fascist living. They take the park as a microcosm of the world and explore the material, natural and human-made, that is found within its borders.

To work with extant material is a contemporary methodology. As Douglas Huebler said:

> 'The world is full of objects, more or less interesting; I do not wish to add any more.' This eco-sensitive model acknowledges that there is enough material in the world and by working with extant material – by selecting it and reframing it – the artist is able to generate new meaning and, in so doing, disrupt the existing order of things.' (Huebler, 1969)

In language, the word 'folly' moves from the French '*folie*' from 'madness' (Old French) to 'delight' (modern French). When I look at this mobile shed bursting with found objects, I cannot help but feel that Williams and son created a folly within a folly when they chose to site this work in the rotunda of a well-known local folly, the Ashton Memorial. Like the shift in meaning from the past to the present, the Williams' folly contains both definitions of the word: an irrational collection that is also a delight. As Sol LeWitt tells us in one of his sentences on conceptual art:

> 'Irrational thoughts should be followed absolutely and logically.' (LeWitt, 1992:837–839).

Williams and son's work operates on the edge of meaning. Their collection fluctuates between sense and nonsense; it is a poetic engagement with the world around us. They have created a phantasmagoria, a dreamscape woven out of the refuse of life, natural and human-made, objects pinned, labelled and collated for our consumption. It is not just the objects but how they are displayed that matters.

Certainly there is a level of conviction in the framing of the work for public reception. And the force of the framing is important. Almost everyone at some point in their lives collects found objects: conkers in the fall, pebbles from the beach, sticks that look like ray guns. We bring them home and arrange them on our shelves with no particular motivation other than sheer fascination with the world we inhabit.

What Robert and his son Jack have done is far more engaged and rigorous. They have chosen a field, they have specified a period and they have then given it the level of attention of a serious scientific research project. Robert Williams and his son Jack masquerade as scientists. This force of intention is what separates the project from an everyday encounter on the nature trail. One could be forgiven for asking where exactly the art is in this earnest collection of the everyday, but the fact that there is so little distinction between life and art is what makes it art. As the French artist Christian Boltanski notes:

> 'If you want to move people, it's always better to be just at the edge
> of life and art... if you want to touch people, it's always better if the
> people who are reading you – the people who are looking at your art
> – don't know exactly if it is art or life.' (Bragg & Fox, 1996)

By working collaboratively, Robert Williams and his son Jack are able to make something more than the sum of their individual contributions. Brion Gysin and William Burroughs refer to this phenomenon as 'the third mind' in their collaborative book of the same name. In their publication, they work to finish each other's 'cut up' sentences in order to make one complete sentence:

> (BG) It says that when you put two minds together...
>
> (WB) ... there is always a third mind...
>
> (BG) ... a third and superior mind...
>
> (WB) ... as an unseen collaborator. (Burroughs & Gysin, 1978:19)

Since the 1960s, artists have recognised the potential of conversation as a practice-based model in its own right. The conceptual artist Terry Atkinson (who taught Williams during his time at Leeds University in the early 1990s) explains the basis of this methodology:

> 'The communal studios, refectories, seminar rooms, vestibules,
> corridors, libraries of the art schools was where the work got made
> and it was generally made in conversation and writing – mostly in
> conversation.' (Atkinson, 1999:12)

In *Thesaurus Scienta Lancastriae*, the objects in the collection are animated by the conversation between father and son. This work is not primarily about examining representational systems or putting them on display (as Dion's practice is); rather, it is about the relation between relations articulated through conversation, through drawing, through writing, and through their collective framing of the work. The conversation between father and son is an integral part of this work.

HOVERFLIES

Hoverflies are the only flies I don't kill. They are only sick on plants, not like poo-flies that are sick on poo. They (hoverflies) are sick on them (plants) to suck the syrup up. A hoverfly has one of these sick things (mouthparts) which can be sick. You can tell it's a hoverfly because of his red eyes. He can't see very well, but he can see the window. He can see light with his red eyes. He likes having his wings lifted up so you can stroke his back.

Hoverflies are the cleverest insect in the world, because they can disguise themselves as parasites. Honeybees and wasps are parasite to catch insects, and so is a honeybee.

HOVERFLIES AND BEES

You can tell a hoverfly from a bee because bees are bigger than hoverflies, and they are hairy – hoverflies have smooth skin, bees have hairy skin. Most of all they can hover, bees can't hover.

Two contemporary artists, Pavel Büchler and Nick Thurston, have paid careful attention to this notion of making through conversation, recently publishing their collaborative text *Word for Word*. In their conversation about conversation, they detail the collaborative space of exchange that occurs in this form of making:

> 'I have told you before that I believe in conversation as a fast, constantly refreshing forum capable of utilising the diverse knowledge and insights of those involved. It is a contributive thing. But what exactly is being contributed? Two or more people are bringing 'something' into a situation, but everything that is brought into that situation is being constantly transformed. So there's a residual effect as much as a noticeable affect… The word 'residual' is interesting here. In conversation the space between conversants is the space in which your contribution changes with every subsequent response, with every new idea that is thrown in. As in cooking, it all boils down to something. The spoken, like the performed, and unlike the written, is 'bound by time'. So, pinning down the affects or effects of your contribution is quite difficult, in the same way that you can't be sure what the carrots have done to everything else in the pot…it's like a chemical reaction. One thought, one statement or any other 'ingredient' may become the catalyst for something that transforms the previous or subsequent statements. They're mutually affecting. They bounce off one another and they affect one another's relative position. They can do all kinds of things. They may 'bounce off' in the sense that they never 'meet again'. They can take off in their separate directions or they may merge together and find some new trajectory or whatever. But it's more than just an intertwining or formation of a texture. It's really like a meltdown in which the two voices, while still perhaps distinguishable via a subsequent reading (listening to the record of that conversation), do not necessarily suggest two individual passages.' (Thurston & Büchler, 2004:55–63)

This wonderful idiosyncratic collection has been built through Robert's and Jack's engagement with the material that surrounds them and through their engagement with each other. Something magical occurs in a collaboration and it is activated by dialogue. Conversation is central to this project, as Robert makes clear when he talks about his work with Jack:

> 'There's something about engaging with him, and his descriptions, and his conversations, and my attempts at interpreting that, which is absolutely fundamental to what the project is about.' (Conversation between the author and Robert Williams, June 2005)

By masquerading as scientists and building their collection across the four seasons, Robert and Jack do question systems of representation, but this is not, for me, the real purpose of this project. The relationship between father and son is the central concern of this project and reminds me very much of a celebrated photograph taken by a father of his son. The image is more than just a portrait of an individual – it clearly documents the special relationship between the two.

SLITHERING AND GLIDING
Slithering is what a snail or a slug does.
Gliding is what a worm or a leech does.

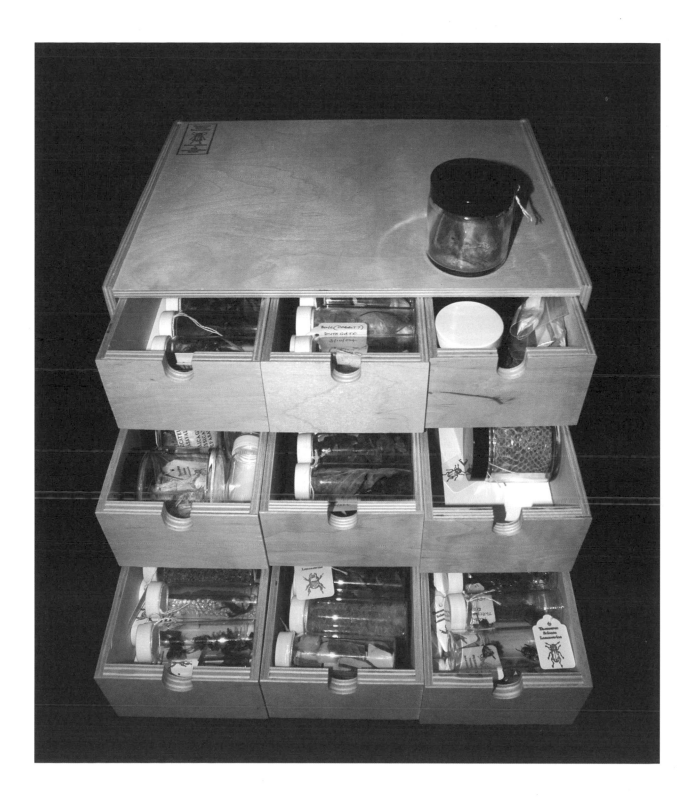

SLUGS

Slugs are like leeches, they get bigger (fatter) when they get smaller (thinner). Slugs eat lettuce leaves with their *radula*. A *radula* is a snail or a slug's tongue, with teeth on it. The teeth are very soft, not like ours. They don't have a mouth, so they breathe through their nostril or air-hole. They stuck their *radula* out of their slime because they don't have mouths.

SLUGS AND SNAILS

Baby snails, when they grow up, lose their shells. Then they find empty snail shells, and move in. Baby snails who don't find shells become slugs.

The image is the American Edward Weston's well-known photograph, taken in 1925, of his son Neil on the verge of adolescence. It is a striking image. One could suggest that it has itself become a dialectical image through its appropriation by the conceptual artist Sherrie Levine in 1979, but that particular controversy is not the reason for my referring to it here. Weston's photograph of his son Neil is both a testament to his craftsmanship – his technical facility – and a record of their intimate father-son relationship. We are lucky to have a written record of Weston's thoughts about this work in his diaries. He makes direct reference to this image:

> 'Beside's Neil's companionship ... he afforded me a visual beauty which I recorded in a series of Graflex negatives of considerable value. He was anxious to pose for me, but it was never a 'pose', he was absolutely natural and unconscious in front of the camera. When I return he may be spoiled, if not bodily changed, in mental attitude. Last Spring in San Francisco at eight years, he was in the flower of unawakened years before adolescence: tall for his years, delicately moulded, with reedlike flow of unbroken line; rare grey eyes, ingenious, dreaming, and a crown of silken blond hair. He is a lovely child!' (Newhall, 1971)

Weston's photograph demonstrates a sophisticated structural arrangement of form and space, and the accentuated lighting heightens the sculptural nature of the pose. Considering the cameras available in the 1920s, the photograph is a technical accomplishment, when even the breathing of the subject could have spoilt the line. In Weston's words:

> 'These simplified forms that I search for in the nude body are not easy to find, nor record when I do find them. There is an element of chance in the body assuming an important movement: then there is the difficulty in focusing close up with a sixteen-inch lens: and finally the possibility of movement in an exposure of from 20 seconds to two minutes – even the breathing will spoil the line.' (Newhall, 1973)

In conversation in June 2005, Robert Williams told me how his son Jack was the child he always wanted to be. *Thesaurus Scienta Lancastriae* is not just an attempt to reawaken history through the dialectical relations created by juxtaposing fragments of the past. Williams and son are using the project to locate their own place within the world. The dialectical image is put into play not just between the objects collected – their own particular history and their recontextualisation within the present – but between Robert's childhood and his son Jack's as well.

As I said at the beginning, the collection always has a double function. Not only does it contain a body of material, but it also affords us an image of the people who select the objects, clean them, preserve them, and put them on display – the collectors. *Thesaurus Scienta Lancastriae*, like Weston's photograph, may be a record of the relationship between father and son, but the image that emerges is an altogether different one. In *Thesaurus Scienta Lancastriae*, it is the father who holds his breath, trying to capture an image of himself through his son.

References

Atkinson T (1999) *Cultural Instrument*. Stoke-on-Trent: Staffordshire University.

Bragg M (ed) Fox G (dir) (1996) *Christian Boltanski* (video). London: Phaidon Press.

Burroughs WS, Gysin B (1978) *The Third Mind*. New York: The Viking Press.

Coles A (ed) (1999) de-, dis-, ex- Volume 3: The Optic of Walter Benjamin. London: Black Dog Publishing.

Corrin LG, Kwon M, Bryson N, Berger J (1997) *Mark Dion*. London: Phaidon Press.

Dion M (2003) in conversation with Zina Davis and Chris Horton, *Collaborations: Mark Dion* Hartford CT: Joseloff Gallery, Hartford Art School, 3–4.

Dion M (2005) Fax from Mark Dion to Simon Morris, 1st September 2005.

Foster H, Krauss, R, Bois YA, Buchloh BD (2004) *Art Since 1900*. London: Thames & Hudson.

Huebler D (1969) In Allberro A, Norvell P (2001) *Recording Conceptual Art*. Berkeley, CA: University of California Press.

LeWitt S (1969) Sentences on Conceptual Art. In: Harrison C, Wood P (eds) (1992) *Art in Theory, 1900–1990: An Anthology of Changing Ideas*. Oxford: Blackwell.

Meyer J (1993) *What Happened to the Institutional Critique?* New York. American Fine Arts/ Colin De Land Fine Art.

Newhall N (1971) Edward Weston: The Flame of Recognition. New York: Aperture.

Newhall N (1973) *The Daybooks of Edward Weston*, Vol 2, *California*. New York: Aperture.

Perloff M (2003) A conversation with Kenneth Goldsmith. *Jacket* 21 (February 2003). www.jacketmagazine.com/21/perl-gold-iv.html

Rendell J (2002) Travelling the Distance/Encountering the Other. In Blamey D (ed) *Here, There, Elsewhere: Dialogues on Location and Mobility*. London: Open Editions.

Thurston N, Büchler P (2004) 'Word for Word', On the Page. *Performance Research:* 9(2): 55–63.

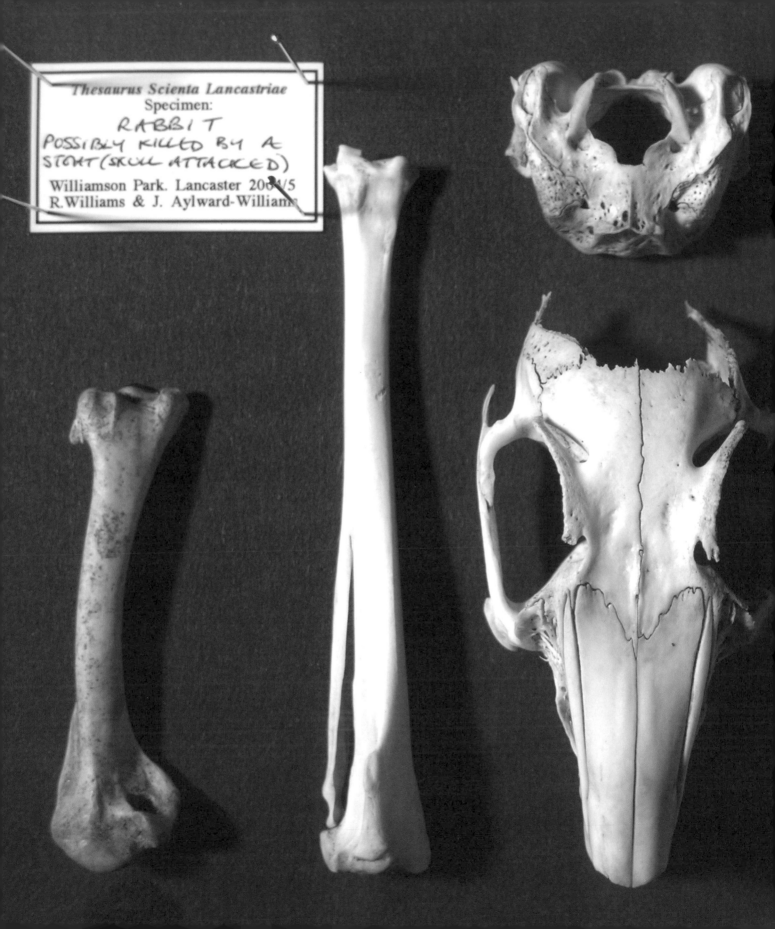

Thesaurus Scienta Lancastriae
Specimen:
RABBIT
POSSIBLY KILLED BY A
STOAT (SKULL ATTACKED)
Williamson Park. Lancaster 2001/5
R. Williams & J. Aylward-Williams

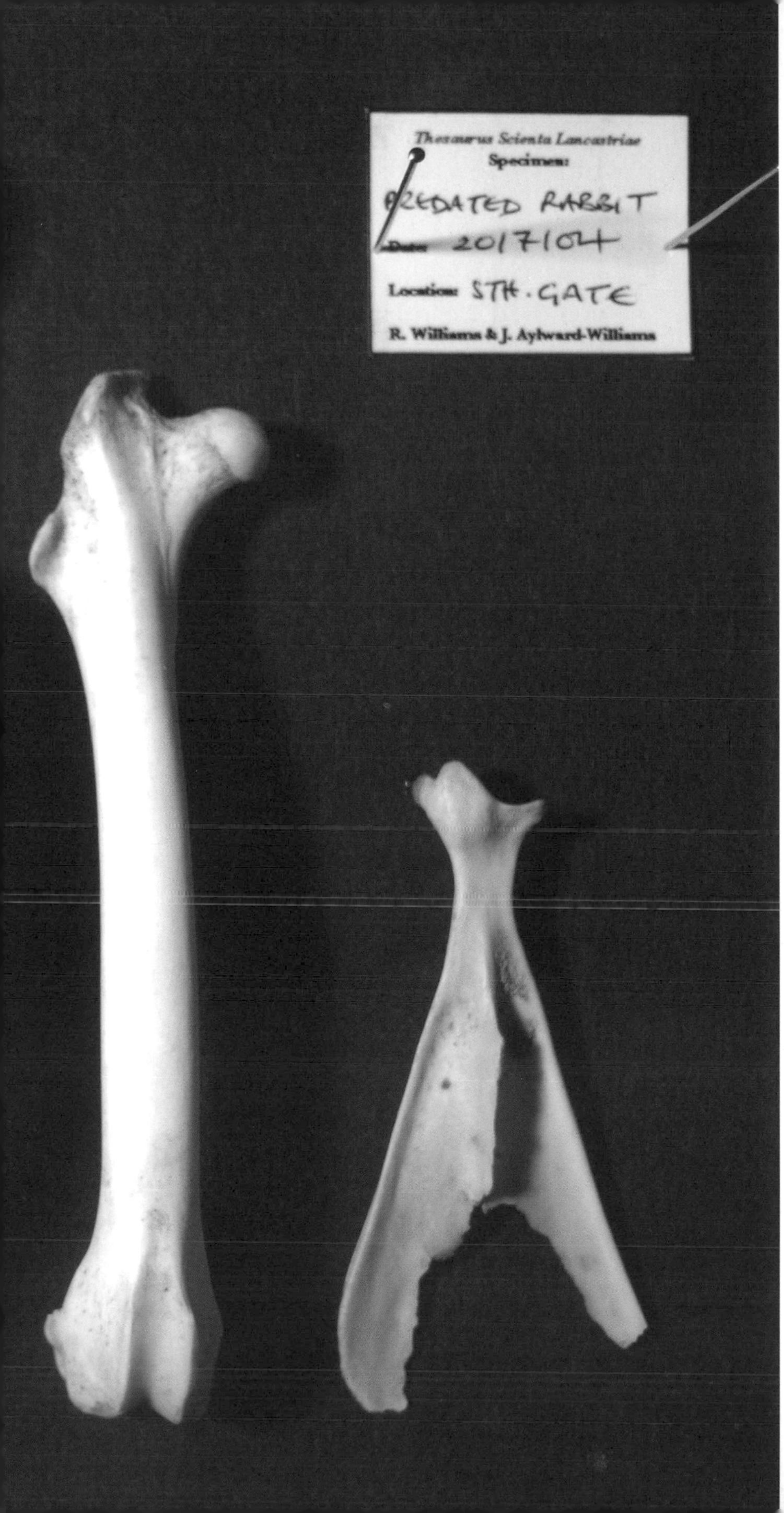

Thesaurus Scienta Lancastriae
Specimen:

PREDATED RABBIT

Date: 2017104

Location: STH . GATE

R. Williams & J. Aylward-Williams

Chapter 2. FINDERS KEEPERS

Reverend Professor John Rodwell

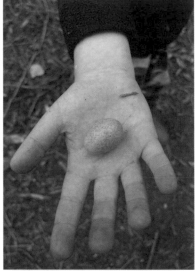

Leap upon curving leap, as much an impression of speed and power as a creature, it had sprung from the still-standing stretch of wheat and, running in a great arc, turned tightly in front of the tractor, just ahead of its advancing wheels, and then was suddenly gone, over the brow.

I was a boy of ten then, and I have spent my life since that time trying to give a name and meaning to such encounters as happened there, as I watched the reaper-binder clattering through the harvest in a golden haze of chaff on a hot summer's day in the 1950s and it set a hare running. That quest to comprehend has one way and another determined my own career as an ecologist and driven a continuing fascination with the natural world. I am loath to give the term 'amateur' to this more general enthusiasm, if it makes it sound amateurish and a contrast to the professionalism of science.

In French, my wife reminds me – languages being her own profession and enthusiasm – an *amateur* is a connoisseur. In both these languages the term really means someone who does something for love, an *amator* in Latin, and then we are in the realm not of payment for work but of things gained and given for the sheer pleasure of learning. Naming is not only to know more, but also somehow to be known and to pay a price for such intimate knowledge. In any case, quite what is professional and what is amateur in the naming and taxonomy of nature is one of the important questions raised for us all by Robert Williams and Jack Aylward-Williams in *Thesaurus Scienta Lancastriae.*

The people who first helped *me* give name and shape to creatures like that hare were certainly themselves amateurs, the enthusiasts and connoisseurs of the Barnsley Naturalist and Scientific Society. That society began, like its counterpart in Lancaster, in the 19th century, though among the less well-to-do in South Yorkshire than here. Five young working men resolved, on 2 February 1867 in George Fogg's barber's shop in Barnsley, to subscribe 'half-a-crown entrance and sixpence per month thereafter in advance' to belong among those who wished to dedicate themselves to the 'observation and study of the objects and phenomena of natural history, geology, mineralogy, chemistry and meteorology' (Atkinson, 1987:8–13).

In my own time among them, they met still – clerks, doctors, shopkeepers, housewives, miners, teachers – every fortnight through the winter, our tread echoing up the stone stairs of the Harvey Institute in Barnsley, to listen to talks by members or from the occasional eminent visiting speaker from the nearby universities of Sheffield and Leeds. Slides were sometimes shown, specimens recently found carefully handed round among us, reports of observations made, experiences shared and tea drunk. And, if I arrived early by bus from my home in the next town or could stay on a little afterwards, there would be a chance to look at the society's museum in the dimly-lit

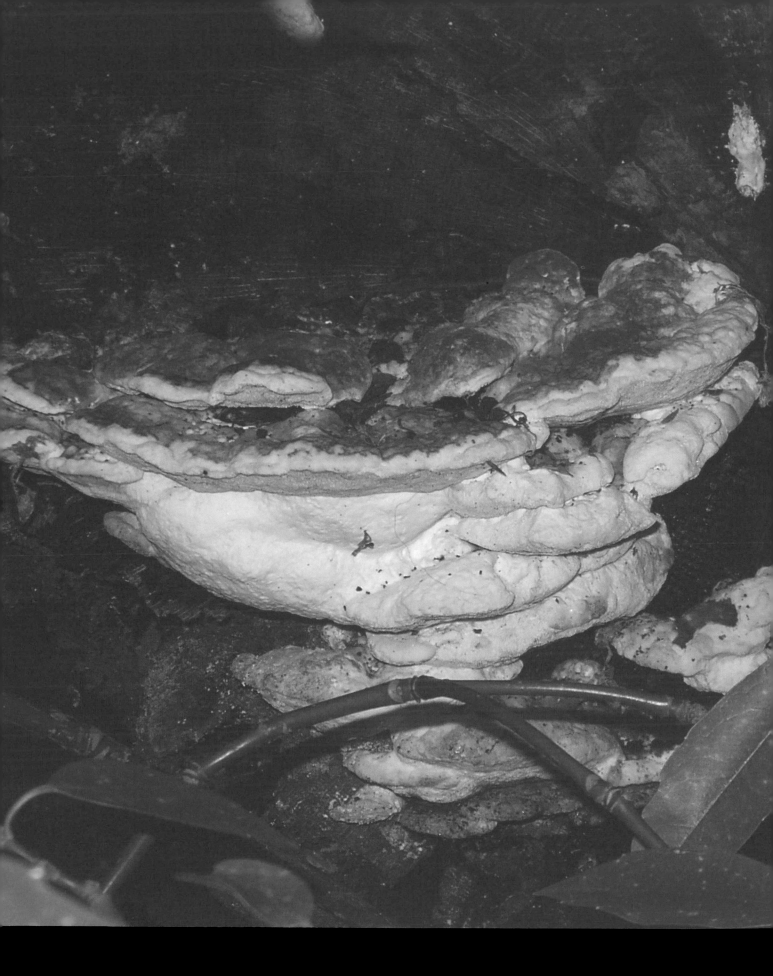

room next door. There, undisturbed by the muffled chatter that came through the wall, along the narrow passageways between high cabinets, I could explore displays of stuffed birds and fish, fossils, shells, butterflies, moths and beetles, skeletons, a crocodile skin. These were testimony to a passion for collecting, curating and organising all that could be gathered by Victorian industry and ingenuity, each item neatly named and labelled, all carefully organised on shelves and trays. For all its now unfashionable deadness, this great cabinet of curiosities quickened in me an appreciation of the sheer exuberant diversity of living nature and the order that might be made of it.[i]

And, from Easter to autumn each year, whatever the weather, the members gathered outdoors every other Saturday afternoon for their excursions to places that were no more than a bus journey away. There, among the collieries and between the tangle of railway lines that made up that part of South Yorkshire then, there was countryside still rich, I was to be shown, in wild animals and plants. With the Barnsley Naturalists, I started to look and learn how to name the things that we saw, calling individual creatures out of the fabric of nature. I liked especially the scientific names of the plants and animals that I was hearing then spoken for the first time. They lent a serious matter-of-fact-ness to our ramblings but had a rich sound of mystery about them as they rolled off the tongue: *Veronica beccabunga, Geranium robertianum, Daldinia concentrica.* And, later those evenings, back home, as we sat around the kitchen table, the rhythm of this recitation would weave itself in and out of those roll-calls of promise from the football results and the shipping forecast on the wireless: Stenhousemuir, Partick Thistle, Queen of the South, Viking, Faeroes, Fisher, Dogger, German Bight.[ii]

My father usually came with me on these trips. On the Saturday mornings, my mother would have made us our sandwiches, always boiled ham with HP Sauce, a combination that I cannot taste now without conjuring from my store-cupboard of flavours those happy days together. We would pack our lunch in Dad's World War II gas-mask bag and, on the photographs which I took of those Barnsley Naturalists' excursions with my Kodak Brownie 127, I see now that almost every other boy carries just such a bag, and that several other fathers stand behind their sons. Apprenticed already then to my dad's wartime skill at aircraft recognition – though just an artillery corporal, he had taught women anti-aircraft spotters this talent – I could myself have told you the identifying features of the Spitfire, the Hurricane, the Typhoon, the Heinkel 113, the Messerschmidt 109, how fighters differed from bombers, what the distinction was between seaplanes and flying boats.

So, in those days, I began to memorise, as if my *own* life depended on it, the differences between the things whose names I had begun to learn and how they were organised into groups – the cranesbills and the storksbills, the finches, warblers and tits, the mayflies and the caddis flies. I made notes, I drew pictures, and for the mushrooms and toadstools, which became a new fascination for me as my first autumn as a serious naturalist arrived, I made a little filing cabinet from empty matchboxes glued together, each drawer filled with tiny cards on which I wrote a name and description of the kinds I might find.

And, in a very real sense, of course, my own life did depend on all this. What school or university or work can do is add detail and shape to the naming and organising of the natural world, but what matters first is the curiosity with which the cabinets of

[i]

Sadly, in a renovation of its premises in the Harvey Institute, many of the Barnsley Naturalists' collections, by then less prized in a world of changing values, disappeared or were disposed of on the local authority tip.

[ii]

Seamus Heaney (2002) has spoken about the beautiful sprung rhythm of these childhood radio litanies in his 'Feelings into Words' from *Finders Keepers, Selected Prose 1971–2000.* In the spirit of 'Finders Keepers', I have borrowed this playground phrase as my own title. Iona and Peter Opie (1959:136–137) provide a summary of how this resonant expression has a continuing currency.

iii

See Wilfrid Blunt (2002), *Linnaeus: The Compleat Naturalist* though it was not, in fact, Linnaeus who gave authoritative form to the classification of fungi, but Elias Fries, another Swede (1794–1878).

GRASS AND DANDELION

Do you know why grass seeds don't have these [dandelion flights]? Because grass seeds are hard, and dandelion seeds are soft. So grass seeds are too heavy to blow in the wind.

encounters and objects we make for ourselves are assembled. Curious children of my own generation were fortunate in that the secondary and tertiary education in natural science still brought you face to face then with many individual creatures and taught how taxonomy and systematics could help comprehend and value their diversity.

Nowadays, though biodiversity is a very popular notion, and television abounds with astonishing testimony to the intricacy and wonder of nature, real animals and plants can be elusive within the landscape of education. The science of systematics is in a crisis, with a dire shortage of skills and practitioners who know how to look, name and organise (House of Lords, 1992, 2002). In the busy little figure of Jack Aylward-Williams, we see that curiosity at work though we are not doing justice these days to the challenge for learning that he poses.

'Giz a look,' my friend Stanley shouted as he jumped down from the tree where we hid to watch people without their knowing. 'What's tha got there? Giz a look.' I reached into my pocket where I had hidden what he had caught sight of and showed him: 'It's a fungus,' I said. 'Looks like a black man's bollock,' he replied. 'It's a fungus.' 'Aye, tha said. What's it called?' '*Daldinia concentrica*,' I told him. 'What sort o' name's that?' I broke open the strangely light fruiting body and showed him the concentric silvery rings of spore cavities inside. 'Where's tha find it?' 'In a wood near Silkstone,' I said. 'That's 'other side o' Barnsley. What's *tha* doin' *there?*'

What I was doing there was discovering another world and a different language with which I could label and organise my own cabinet of curiosities – things like that bizarre fungus that I had been shown on the dead branches of ash trees in the wood near Silkstone. I was learning that every creature has two names in what is called a Latin (or latinised) binomial – an epithet unique to the species (*concentrica* in this particular case) and the name of the genus it belongs to, along with other species of a closely related kind (*Daldinia* here). Such binomials effectively summarise more cumbersome descriptions of the characteristics of a particular creature, often telling us something about the colour, shape or structural characteristics of the species, something of its behaviour or the habitat where it occurs, an indication of the region or country it is found in. Sometimes it commemorates the person who first described the animal or plant. With this fungus, for example, *concentrica* describes the rings of spore chambers laid down in each year of the fruiting body's expanding growth while *Daldinia* remembers the fungus enthusiast Daldin – though we do not now seem to know who he actually was.

So effective and economic is this format that binomials still provide a universal shorthand for anyone, irrespective of their native tongue, who is interested in talking in a scientific way about the natural world. It was Carl Linnaeus, the son of a Swedish Lutheran clergyman, who most definitively gave these kinds of names their final authoritative form. He provided, in his *Systemae Naturae* of 1758, a survey of all the animals and plants that were then known to the world of science. Linnaeus, too, gave clearer shape to how plants and animals might be organised in classifications of the species and genera into families, orders and classes – hierarchies of nature that help us see the inter-relationships of creatures, one to another, within a systematic frame.iii

In fact, I would learn later that, behind the poker-faced authority of taxonomy and systematics, there lie many misreadings of what plants and animals are really like,

or how they behave, where they come from, or how they relate to other creatures that are more or less similar to themselves. With hares, for example, it was Linnaeus who gave the animal that he himself knew in Sweden, an agile creature that lived a solitary life, the name *Lepus timidus* – the shy leaper. For a hare of which he heard tell from South Africa, an altogether bigger animal lacking the snowy-white coat that its relative acquired in the autumn and winter moult, he coined the name *Lepus capensis*, the leaper from the Cape. What Linnaeus did not know was that *L. timidus* did not actually extend throughout temperate Europe but was just a northern and mountain creature. *L. capensis* did occur throughout lowland and southern Europe and indeed into much of Asia. So, in Britain we have the odd situation that our common lowland hare, the one that bewitched me that day in the corn, was given a name smacking of an exotic southern hemisphere home. And, if anything, we now know that this particular hare is rather more shy than the mountain relative we call *timidus*.

We do have that northern mountain species, too, in the Scottish Highlands and through the whole of Ireland, though there it occurs in a form that is incompletely white in winter or not at all. Taxonomists have therefore recognised this finer distinction at sub-specific level by giving these two creatures the benefit of triple-barrelled scientific names – *L. timidus scoticus* and *L. timidus hibernicus*, the terms acknowledging their respective native homes. And *L. capensis* has recently been renamed *L. europaeus*, an altogether more sensible epithet in the light of what we now know about the distribution of the hares of the world (Harrison Matthews, 1982; MacDonald 1989; Ewart Evans & Thomson, 1972). The former name is now used exclusively for the hare that is really at home in Africa but which comes no further north in western Europe than southern Spain – or so we think at the moment. Science is a provisional affair and there is always room for new discoveries. Even now, it is often educated amateurs who add to our store of new species records, reorganise the way in which science names and classifies them and help us understand the behaviour of animals and plants. Jack is therefore himself working at this forefront of exploration.

Meanwhile, the language my friend Stanley used to talk about the fungus I showed him has a much wider commonality than the scientific way of speaking which became my own trade. Stanley spoke then of what he saw, what he fancied the thing I showed him reminded him of, and indeed one of the most frequent vernacular names for *Daldinia concentrica*, is 'Carbon Balls'. Then, there's 'Coal Fungus' or, in a more historical vein, 'King Alfred's Cakes', this last recalling that schoolroom story we had been told about how, sheltering in a poor home after a battle and deeply preoccupied with the fate of his kingdom, Alfred had let the cakes on the fire in front of him burn. The look of a thing, the visual associations it conjures up – these are often the basis of such vernacular names for plants and animals. For example, the time that a plant flowers or an animal appears to mate (as with May-flower for *Crataegus monogyna* and May-fly for the invertebrates we group together in the Ephemoptera), or the habitats they occupy (Wood Anemone for *Anemone nemorosa*, House Sparrow for *Passer domesticus*). Often, however, vernacular names may have some more allusive (and elusive) link with folk knowledge.

Specimen: Jack-by-the-Hedge
Date: 1/505
Location: Williamson Park
Robert Williams & Jack Aylward-Williams

HOVERFLIES DISGUISED AS HONEYBEES

They [hoverflies] try to disguise themselves as honeybees, but people really do know they are hoverflies. They disguise themselves as honeybees so they don't get hurt. They are like ones that disguise themselves as wasps, the rare ones. At the moment they are quite rare, but they are getting common.

LADYBIRDS AND BEETLES

Ladybirds are fly-beetles. The headless beetle
(found in Williamson Park) in the woods is
a beetle-beetle. Cockroaches are cicada-
beetles, the ones that have wings. Hissing
cockroaches are giant woodlice. A cockroach
is like a ladybird because a cockroach turns
into a cicada, and a ladybird turns into a fly.
A headless beetle, if it had been alive, would
have turned into a bee because it would have
grown some hairs. Wing cases sometimes
fall off ladybirds when they are dead.
Especially those that are as small as ants.

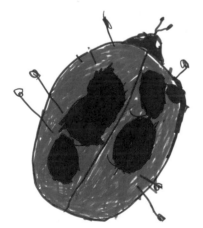

Geranium robertianum for example, is one of the Cranesbills, so called because of the similarity of their fruits to the bird's sharply pointed beak (and geranion which has got latinised to *Geranium* is Greek for crane too). So we have 'Wood Cranesbill' (for *Geranium sylvaticum* from woodland glades) and 'Bloody Cranesbill' (because of the bright red flowers of *Geranium sanguineum*). However, this particular plant of woodlands and hedge-bottoms whose scientific name was among the first that I learned, is commonly called 'Herb Robert'. It would be a boring suggestion that its specific epithet is a corruption of *ruber* reflecting the attractive reddish tinge of the stems and the name is generally supposed to invoke Robert, Rupert or Robin, as in Robin Goodfellow, the impish creature of folklore, though quite why is unclear: maybe because of the uneasy smell of the plant. Its rarer name of 'Fox Geranium' may likewise come from the German *Volks-* for fairy folk and the particular reason for such names is often now lost in time or among a mass of whimsy. Sometimes, too, linguistic tangles lie in both the Latin and different vernaculars. With *Veronica beccabunga*, for example, another of those early plants I learned to identify, the specific epithet may be a corruption of *bachbunge*, German for 'stream bunch', which well describes its growth form, or maybe the Flemish *beckpunge* for 'mouth smart' which tells us how bitter it tastes. Its vernacular name 'Brooklime' refers to the muddy habitat in which it commonly grows.

Others take the common name 'Herb Robert' as making a link between St Rudbert, a saint of Salzburg who had a reputation for healing wounds, and the use of an extract of *Geranium robertianum* for supposedly reducing inflammation. This medicinal origin of vernacular plant names is very widespread and often reflects some aspect of the plant's appearance or smell that is supposed to be a tell-tale indication of its particular value – the doctrine of signatures as herbalists called it, which has given us common names 'Woundwort', 'Lungwort', 'Nipplewort', 'Liverwort' and so on. It is the knowledge of herbalists, often arcane but sometimes full of wisdom, from which much of our more modern science of botany has come. Meanwhile, whether *Daldinia concentrica* can actually combat cramp when carried in the pocket, as another of its names 'Cramp Balls' suggests, seems unproven. Myself, I had only wanted to take the fungus safely home for my bedroom shelf.

LADYBIRDS

Ladybirds change to hoverflies. Underneath
the eyes are things that look like hairs (these)
are really a timer, (which goes off) and then
the wing cases open and drop off when the
face drops off. Then it grows again there
will be no nose, no mouth, but a sick-thing
(*proboscis*) to be sick on food, so a fly can eat
it, then the eyes can go from white to red.
The fly can't see very well, but it can see
light and has a great sense of smell.
Ladybirds are hoverfly larvae. Rain beetles
are poo-fly larvae. The beetle we found with
no head used to be a bee larva.

iv
See, for example, the tongue in cheek remarks of Humphrey Gilbert-Carter in the Foreword by Max Walters to Dony et al (1974). A more biting attitude is taken by Alan Mitchell (1974) in the Preface to his (nonetheless excellent) A Field Guide to the Trees of Britain and Northern Europe.

v
The best single book to open up this treasure house of names and meanings among the plants of this country is Flora Britannica by Richard Mabey (1996). More recently, Birds Britannica (Cocker & Mabey 2005) provides a companion volume. An immensely enjoyable exploration of the way in which nature enriches our common language can be found in Pedigree: Words from Nature (Potter & Sargent, 1973).

Such vernacular names for particular creatures can be very numerous and localised, representing the rich diversity of ways of looking and speaking in different parts of the country. For *Arum maculatum*, for example, a plant of woods, hedgebanks and wilder shady gardens, there are more than 90 names (Grigson, 1955; Prime, 1960), many of them with a knowing vulgar take on its peculiar inflorescence, a soft green spathe with a protruding purple spadix which unfolds among the glossy green leaves in late spring. Thus it is that we have 'Lords and Ladies', 'Foolish Lovers', 'Cuckoo Pint' (short for 'pintle' or penis) or, more obviously, 'Priest's Pilly'. 'Willy Lily', or 'Dog's Cock'.

To scientists, busy inside their own cabinets of curiosities – ordered, objective, rational, so we think – this can be all too common for words.[iv] Yet, despite the disintegration of shared folk beliefs and religious practice, and the erosion of the local communities in which such language and learning often grew, vernacular relationships with nature remain surprisingly strong and we ought not to underestimate their power to give meaning to where we live, what we do and who we are.[v]

Nourishing and organising our imagination in relation to nature is what *Thesaurus Scienta Lancastriae* is about. The project can help us see that no-one will necessarily take us to task if we wish to celebrate our shared life with nature with common talk, idiosyncratic classifications, offbeat trains of thought or flights of fancy. It would be hard to imagine that there was not some sheer pleasure in the way entomologists have conjured up such uncommonly beautiful names as the Processionary, the Saxon, the Rosy Footman, the Lappet, the Smoky Wainscot, the Merveille du Jour, simply

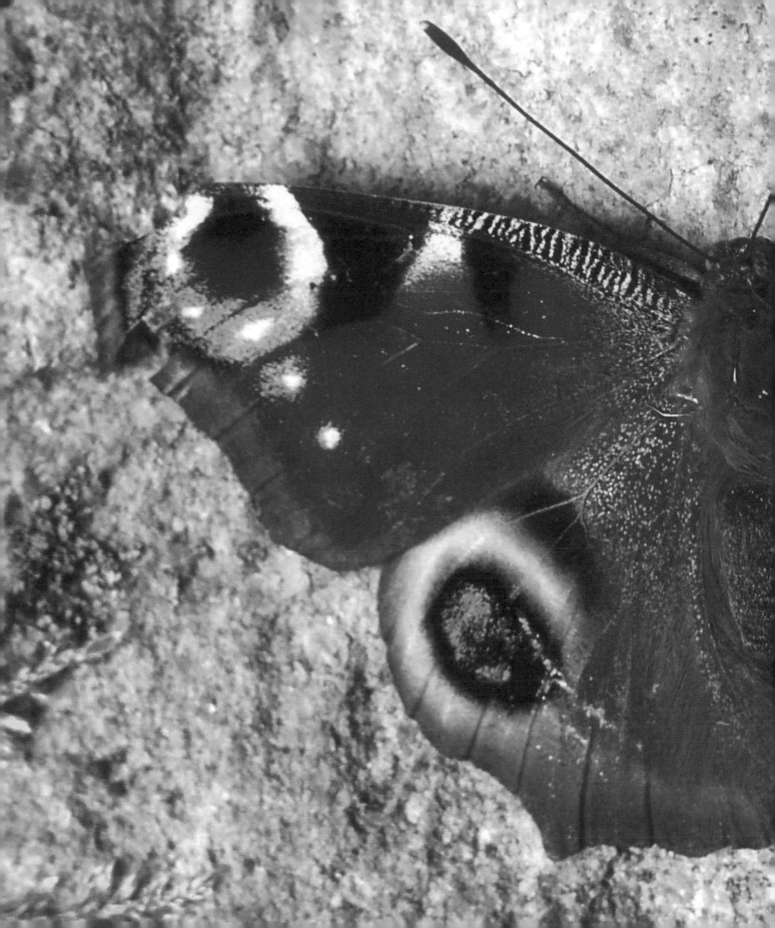

GREAT POND SNAIL

That thing is its horn – the thing it is putting in because it doesn't want to see. Sometimes they lay silly eggs on pond weed, that are all squidgy, and that when you pick them up, it turns into snot.

to celebrate the mothiness of moths. The books in which I first encountered these wonderful names were the two volumes of *The Moths of the British Isles* by Richard South (1907). Recently marooned by bad weather on Skomer Island, I found these volumes again, idly left on the research station desk by another visitor. After over 40 years, the names still worked their magic.

Deeper inside the cabinets of curiosities that are our own minds and hearts, behind the door where the chatter of science and common talk are muffled, we revisit, sort and rearrange the things we have found and kept, discovering again lost treasures and something of ourselves in them. Here, for example, among my own belongings, I could show you a picture of an ant, a photograph of a young woman and two dirty sheets of glass. The ant is *Lasius niger*, the Common Black Ant, indeed the most widespread ant that we can find in Britain, often nesting close by, under the pavements outside our own homes and in our gardens but, for me, a far from commonplace creature. I was shown it first when a lovely student teacher came to my primary school for just two weeks bringing with her a nest of these ants in a formicarium. Each morning she let me into school early so that together we could feed the ants with a dollop of sugar syrup. I would watch them bewitched as they scurried along the passageways they had excavated in the thin layer of soil sandwiched between the two sheets of glass. For weeks afterwards, I would spend many hours sitting in the sunshine outside my home, watching the ants coming and going in and out of the cracks between the paving stones, thinking how I might capture a colony for the formicarium which my father was making for me in his shed.

Then, here I have a plant fossil with delicate leaves on a sliver of shale, a letter from Vienna and a hammer. The fossil is *Neuropteris*, a name that is given to fragments of fern like leaves that grew on trees, the ancestors of the cycads we can see these days for sale in garden centres but which, in the Carboniferous period, dominated in tropical swamps that became compressed into coal over the space of 220 million years. I first saw this name *Neuropteris* written out in a striking script, on a label fastened to a bundle of newspaper handed to me by a dashing young Austrian mining engineer. Coming on a placement to the colliery where my father worked as a wages clerk and befriended by him, he brought for me, each carefully wrapped, minerals and fossils that he had prised from the seams deep below our town. There were big, shiny crystals of galena, an ironstone nodule heavy in my hand and, best of all, *Neuropteris* leaves impressed as fragile black wafers of coal on to shale. Afterwards, my father had a geological hammer made for me in the colliery blacksmith's shop and we went geologising for rocks ourselves.

Next among my curiosities, there is a skull of *Halichoerus grypus*, the Atlantic Grey Seal, a seaside postcard from Llandudno and a book *Nature is my Hobby* (Adams, 1958). The seal breeds on the rocky shores and islands around our coasts, though the first one I saw myself was dead, washed up below the cliffs where I was walking on a seaside trip. Its weighty body was bigger than me, blotchy and bloated. An intrepid naturalist by then, I cut through the rotting flesh of its neck and carried off the head, letting it stew in one of my mother's buckets for some weeks and then scrubbing the skull bone-white with bleach. The Christmas before, the Secretary of the Barnsley

POND-SKATERS

Pond-skaters can jump and waterskate. They eat lots of things in the water, like fish and mud. Pond-skaters are very hard to catch in a net because they move fast, and I think they can fly.

Naturalists had given me Adams' book of wondrous practical advice such as how to stuff a dragonfly, how to correctly mount and label dead beetles on to cards, and, what had most beguiled me most, how to clean and preserve skulls. The seal's head became the prize exhibit in my bedroom museum.

Joan Cooper the student teacher, Alfred Weiss the mining engineer, and Ralph Atkinson the clerk from the education office in Barnsley – these prize friends first shared their own inquisitiveness about nature with me and opened to me a particular cabinet of their own experience and enthusiasms. The strength of my own odd passion for nature somehow gave my parents the confidence to invite these people to tea, an unheard-of notion in the working-class culture in which we all grew up. The garden where we took their photographs, the shed where my father made me things, the bedroom where I had my collections and my books, the pavements outside, the walk to school, those woods and heaths a little further afield which the Barnsley Naturalists visited year in, year out, these were where such people made spectacular creatures familiar to me. Thus it was that nature helped me comprehend where and among whom I belonged, helping people and place know each other.

It was the poet John Clare who spoke not so much of knowing a place as being known by it. Born in Helpston in Northamptonshire in 1793, his very first poem, named after that village, is stamped with this strong sense of attachment and recollection of favourite spots. Taking a walk with Clare you think you are encountering the commonplace but his sharp eye, the way he sees, the ways he describes, makes each thing extraordinary. These insights are from Jonathan Bate's (2003) lovely biography *Clare*. In a world even fuller now, than for Clare, of 'vague unpersonifying things' as he called them in his poem *The Flitting* (1832)[vi] the collections that Jack and Robert Aylward-Williams have assembled for us, finding and keeping in their shed, provide just such a recollection of objects and experiences that might easily escape us as mundane, giving us a sense of belonging, helping us know – and be known by – a place.

Many of us have walked, some of us often, in Williamson Park, taking it in our stride and for granted, yet *Thesaurus Scienta Lancastriae* helps us see it as a very particular place, a microcosm of the wider landscapes of our lives in which we coincide with nature in one way and another. Williamson Park is a place where science, entertainment, exercise and fun have all co-existed. Through its century of historical life, as trees have matured, buildings have been built, renovated or subsided into decay; in its yearly round of plays and exhibitions, sunbathing and sledging, leafing and fall, flowering and fruiting, nesting and fledging; and week by week as dogs are walked, prams pushed, balls kicked around. All these things have left their mark and their memories one way and another, and in and around them Jack and Robert have made their own way, as the fancy of nature has taken them.

In *The Names of the Hare in English*, a 13th century poem, probably written in the Welsh borders (Ross, 1935)[vii] we hear 77 different vernacular names for the creature – among them, 'the frisky one', 'the springer', 'the slink-away', 'the starer with wide eyes', 'Old Big-Bum' and (one specially for me) 'the creature who dwells in the corn'. Quite what this litany is, is not entirely clear but, since many of the names for the animal are teasing or even abusive, it could be a kind of incantation that was recited to bewitch it and deliver it into the hunter's hand. That's quite a particular apprehension

DRAGONFLIES
Red dragonflies are called 'Ruby Red Darters' – green ones are called 'Emerald Darters'. Blue dragonflies are called 'Sky Blue Azurite Darters'.

vi
The Flitting was written after Clare's heartrending move from Helpston to Northborough in 1832, a journey of only three miles but a world away from where he felt he belonged.

vii
A Welsh parallel by Dafydd ap Gwilym, written at the time of Chaucer, abuses a hare which scared off a girl that the poet had arranged to meet.

viii
And, as it happens, a sense of belonging that is distinctively Welsh – *cynefin* in the native tongue. This notion is very nicely discussed by Andrew McNeillie (2005a) and glossed in one of his own poems in *Slower* (2005b) as: 'a sense of being that embraces belonging here and now in the landscape of your birth and death its light and air, and past, at once'.

of nature, though it is one which we have to admit has, in the past, filled many cabinets of curiosity with birds' eggs, beetles, butterflies, and so on, and from which we have actually learned and wondered much about the diversity of nature. Jack's finding and keeping have been different since he resolved to collect only what had fallen and died already. Nature has thus imposed its own bittersweet seasons and serendipity on the *Thesaurus Scienta Lancastriae*, giving it a distinctiveness of time as well as a place, an intersection of the there and then which, wheeled here and there in the shed, gives us meaning here and now.

Of course, in time sheds also disintegrate, collections are lost from our sight and we move on to other places. When my own parents flitted from South Yorkshire, leaving our home, the garden, the field where I had seen the hare, the friends, the shed, it was grudging and tearful that I went with them to north Wales. I watched the soaring fulmars cutting the gusty air off the cliffs of the Great Orme, tracing the sequence of seaweeds that marked the push and pull of the tides down the rocky shore, finding the rare goldilocks and hoary rock-rose on the cliffs. The nature that already seemed to know me provided a kind of welcome and solace of its own. [viii] In Williamson Park, too, we can be sure that we have a place where trysts have been made, hearts broken and wounds bound. Through Jack and Robert's finding and keeping, we may discover ourselves too to be found and kept by nature for another time and place.

References

Adams CVA (1958) *Nature is my Hobby.* Exeter: Wheaton.

Atkinson RS (1987). 'The first hundred years of the Barnsley Naturalist and Scientific Society'. In Lunn, J (ed) *Wildlife in Barnsley*. Barnsley.

Bate J (2003) *John Clare.* London: Picador.

Blunt W (2002) *Linnaeus: The Compleat Naturalist.* Princeton, NJ: Princeton University Press.

Clare J (1832) *The Flitting.* In Thornton, K, Tibble A (eds) (1979) *The Midsummer Cushion.* Manchester: Carcanet.

Cocker M, Mabey R (2005) *Birds Britannica.* London: Chatto & Windus.

Dony JG, Perring F, Rob CM (1974) *English Names of Wild Flowers.* London: Butterworth.

Ewart Evans, G, Thomson D (1972) *The Leaping Hare.* London: Faber and Faber.

Grigson G (1955) *The Englishman's Flora.* London: Phoenix House.

Harrison Matthews L (1982) *Mammals in the British Isles.* London: Collins New Naturalist.

Heaney S (2002) 'Feelings into Words' from *Finders Keepers, Selected Prose 1971–2000.* London: Faber and Faber.

House of Lords (1992) House of Lords Select Committee on Science and Technology. Systematic Biology Research. HL Paper 22-I. London: HMSO.

House of Lords (2002) House of Lords Select Committee on Science and Technology *What on Earth? The Threat to the Science underpinning conservation*. HL Paper 118(i). London: The Stationery Office.

Mabey R (1996) *Flora Britannica*. London: Sinclair Stevenson.

MacDonald D (ed) (1989) *The Encyclopaedia of Mammals*. London: Unwin Hyman.

Mitchell A (1974) A Field Guide to the Trees of Britain and Northern Europe. London: Collins.

Opie I, Opie P (1959) *The Lore and language of Schoolchildren*. Oxford: Oxford University Press.

Potter S, Sargent L (1973) *Pedigree: Words from Nature*. London: Collins New Naturalist.

Prime CT (1960) *Lords and Ladies*. London: Collins New Naturalist.

Ross ASC (1935) *Les nouns de un leure en angleis* in Bodleian Digby 86. *Proceedings of the Leeds Philosophical Society*, Literary and Historical Section 3, 347–77.

South R (1907) *The Moths of the British Isles*. London: Warne Wayside and Woodland Series.

HUMAN BEINGS

Human beings are *reptars*. They are the only *reptars* that do not lay eggs. They have eggs inside them, at least the girl ones. The eggs hatch. Their eggs are called cells. They look like blood cells.

Chapter 3

**A SMALL LAMP OF SCIENCE:
THE LANCASTER SCIENTISTS
AND THE GREG OBSERVATORY**

Peter Wade

Perhaps it was something in the water, but in the late 18th and early 19th centuries Lancaster was home to an unusually high number of scientists, of whom some went on to be ranked among the great and the good of the Victorian age. The four most eminent were: William Whewell (1794–1866), Richard Owen (1804–1892), Edward Frankland (1825–1899) and William Turner (1832–1916). Others from a similar period with Lancaster connections included: mathematician John Dawson, chemists Robert Galloway, George Maule and Edmund Atkinson, civil engineer James Mansergh and electrical engineer Ambrose Fleming.

Most notable of these others was William Sturgeon (1783–1850), a 'poor man's Faraday', who was born in the Lune valley village of Whittington and went on to build, among other electrical devices, the first electromagnet. Even Michael Faraday himself had family connections in the Lancaster area.

Whewell, Owen, Frankland and Turner all grew up in a Lancaster that had enjoyed the wealth derived from its trading links with the West Indies but which had also seen that trade decline. These were the years, too, that textiles and particularly cotton were beginning to become an important element in the economy of the town – though it would be some years before the manufacture of linoleum rose to the fore. As the four future scientists went to school and/or served their apprenticeships, Lancaster was in a sense drawing breath between one period of prosperity and the next, though each of the boys' lives was in some way touched by the Atlantic trade of the 18th century.

William Whewell

Master of Trinity College, Cambridge; Professor in turn of Mineralogy and Moral Philosophy, the Reverend Dr William Whewell's interests spanned 19th century science and a good deal more besides as his writings show, covering mechanics, dynamics, the tides, the history and philosophy of science, moral philosophy, architecture and literature. Yet, had it not been for a chance encounter over a garden fence, the world might have lost an intellectual colossus while Lancaster gained an erudite carpenter. The world might also never have had the word *scientist*, which Whewell coined as the equivalent in the sciences to the umbrella term *artist*. Whewell introduced the word in his book *The Philosophy of the Inductive Sciences, founded upon their History* of 1840 in which he commented:

William Whewell

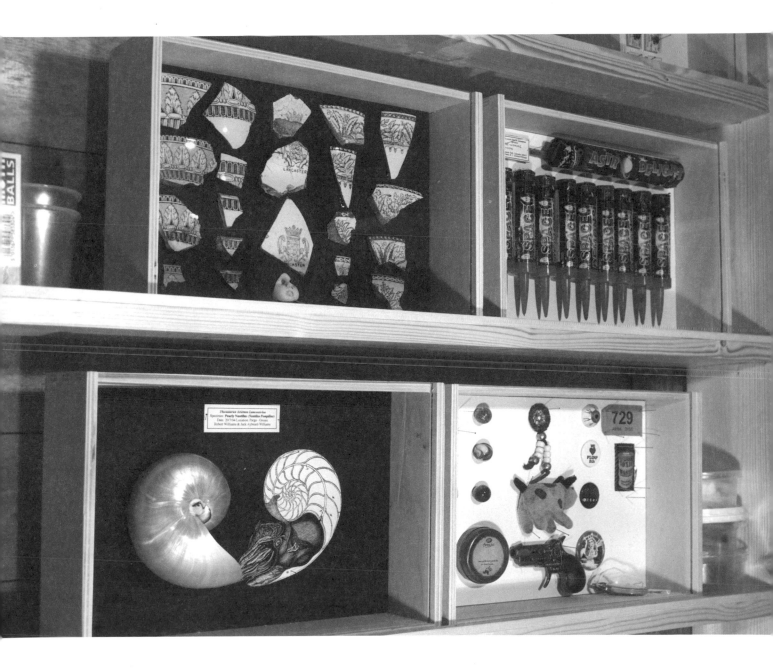

Right: Whewell is introduced to the young
Queen Victoria and Albert, the Prince Consort

THE QUEEN IN THE WOODWARDIAN MUSEUM.

'We need very much a name to describe a cultivator of science in
general. I should incline to call him a scientist.'

Whewell was by then an accomplished wordsmith, having devised geological
terms in 1831 and, in 1834, words to describe Michael Faraday's latest electrical
discoveries. The terms *electrolysis*, *electrolyte*, *electrode*, *anode*, *cathode* and *ion* were all
invented by Whewell at this time.

William Whewell was born in Brock Street, Lancaster and was the oldest of the
seven children of John Whewell (a master carpenter) and Elizabeth (née Bennison).
The Whewells were a well-established Lancaster family. John's trade of carpenter
could well have depended in part on use of the woods imported to Lancaster from
Africa, the Baltic or Canada.

William attended Lancaster's Charity (or Blue Coat) School, which since 1770 had
provided free instruction in the 'three Rs' for the sons of craftsmen. One lunchtime,
recalled another of the Lancaster scientists, Richard Owen, John Whewell was called
upon to mend the fence between the houses of the Owen family and Joseph Rowley,
headmaster of the grammar school. John brought William along with him as an
assistant since he planned to have William apprenticed as a carpenter. Rowley came
down to speak to the pair and was impressed by William's answers to his questions,
especially in arithmetic. As a result, Rowley offered William a free place at the school
and persuaded his hesitant father to set aside plans for his apprenticeship.

The result was a rather odd pair of schoolfellows. On the one hand was Richard Owen who had been allowed to attend the grammar school at only six years of age because Rowley was his godfather. On the other was the 16-year-old William Whewell whom Owen recalled as a 'tall, ungainly youth'.

Not long after, Whewell was examined by John Hudson, vicar of Kendal and a tutor at Trinity College, Cambridge. As a result of their meeting, Hudson predicted that Whewell would be one of the top six Wranglers of his year (i.e. placed among the six best students taking the Mathematics tripos). In the event, Hudson was proved correct although Whewell took second place to Edward Jacob who was the Senior Wrangler of 1816. Jacob, it seems, had given his rivals a false sense of security by assuming a pose of apparent idleness though Whewell also claimed that he had not been able to write fast enough in the final examination.

To reach Cambridge, Whewell had to transfer to Heversham Grammar School so that he could compete for an exhibition worth £50 a year. In the absence of a suitable candidate for the exhibition drawn from among the parishioners of Heversham, Whewell was granted the exhibition in 1811 on condition that he spent two years at the school there. Such was his standing that, upon the death of the headmaster, Whewell at the age of 17 effectively took charge of the school as deputy master until the appointment of a replacement in 1812. He finally began his studies in Cambridge in October 1812.

Though Whewell spent only a short time at the grammar school in Lancaster, while he was there he founded a lifelong friendship with Richard Owen. In 1842, Whewell returned to Lancaster for a public dinner in honour of Whewell himself and Richard Owen.

On arriving in Lancaster itself which had been enlivened by peals of bells rung out from the Priory for the event, Whewell visited the grammar school requesting that the headmaster grant the boys a holiday (a request also made by Owen).

In his speech at the dinner, Whewell noted how the, 'forms of rivers and the early scenes of youth – your castle towers – the waters of the Lune – have haunted me when absent.' Whewell then quoted from William Wordsworth's poem of 1807 *My heart leaps up when I behold* which includes the line 'The Child is father of the Man'. Whewell felt this to be true in his own case, allowing him to trace his mature work back to his childhood experiences. Whewell's later interest in the tides may well relate back to his boyhood experiences on the shores of the tidal River Lune.

Whewell spoke of there being but two scenes in his life, Lancaster and Cambridge, and of the connection between them by which he was:

> '…enabled in mature years to follow out those trains of speculation, those objects of thought, of which the plan and image floated over my early academic life, and shall I say were almost present before me as I rambled on the banks of the Lune.'

RAIN, RIVERS AND WATER
The rain fills the rivers. The river are putting water into the plants. Then, you go and get water from your taps.

Sir Richard Owen and his granddaughter
Right: The site of Owen's birth place
in Lancaster

Whewell died in 1866 following a riding accident and was buried at Trinity College. In Lancaster, a memorial was placed on the southern wall of the Priory. The inscription reads:

> 'Sacred to the memory of William Whewell DD XXIV years fellow and tutor of Trinity College Cambridge then XXIV years master of the college. Born at Lancaster May XXIV MDCCXCIV. Died at the Lodge of Trinity College March VI MDCCCLXVI and was buried in the ante-chapel of the college. This window was erected by his only surviving sister [Ann] as a tribute of affection to the memory of a much loved brother. Thy brother shall rise again.'

Richard Owen

Professor Sir Richard Owen, Hunterian Professor of Comparative Anatomy at the Royal College of Surgeons, Conservator of the Royal College's Hunterian Museum, Superintendent of the Natural History Departments of the British Museum and founder of the Natural History Museum was, like William Whewell, a wordsmith. His contribution to the English language was the word *dinosaur* first used by Owen at the 1841 meeting of the British Association for the Advancement of Science held in Plymouth. Drawn from its Greek roots *deinos* meaning *fearfully great* and *sauros* meaning *crawling animal* or *reptile*, *dinosaur* is usually taken to mean *terrible lizard*.

Fossil reptiles had been known since the 1820s. Among the earliest identified examples were the *Megalosaurus* found by Dean William Buckland, Professor of Geology at Oxford and the *Iguanadon* and *Hylaesaurus*, both found by a medical doctor and amateur geologist, Gideon Algernon Mantell of Lewes in Sussex. They were recognised as being distinct from other fossil creatures by their large size and vertical legs. It was thought that they might have resembled modern animals such as the rhinoceros, hippopotamus or elephant. Mantell wrote to both Professor Buckland and the eminent French anatomist Georges Cuvier about his findings as well as visiting the Hunterian Museum (prior to Owen's period there) where he noted the similarity between the teeth of his fossil animals and the modern iguana.

On New Year's Eve 1853, Richard Owen took part in a dinosaur dinner held inside a partly completed model of an Iguanadon in the grounds of the Crystal Palace. The models had been created between 1852 and 1854 by the sculptor Waterhouse Hawkins in consultation with Owen. The models drew Owen's name into the public arena for the first time and remained his most popularly known works. At the dinner, Owen took pride of place at the head of the table at which about a dozen were able to dine in the somewhat cramped conditions. The model was housed in a tent carrying the names of Buckland, Cuvier, Mantell and Owen.

Owen's work in anatomy won him an influential place in Victorian society. He was described by *The Times* newspaper in 1856 as the most 'distinguished man of science in the country' and as a naturalist was second only to Charles Darwin. Owen could number many of the great and the good of the day among his personal acquaintances. The Prince Consort was attracted by his books and invited him to lecture to the royal children in 1860. He lectured to the royal family at Windsor again in 1864 and was given Sheen Lodge in Richmond Park by Queen Victoria in 1852.

REPTILES AND REPTARS

Reptiles are: snakes; lizards; crocodiles; alligators; geckoes. *Reptars* are: snakes with no tails; lizards that breathe fire; mammals; monsters; gruffles. Dragons are reptile/*reptars* (crossed), so are gruffles. All creatures that breathe fire are reptile/*reptars* (except. of course, fire breathing lizards which are *reptars*). A hydra is a true reptile.

DINOSAURS

A dinosaur is a mammal, and can lay an egg. A mammal is another word for *reptar*. All *reptars* have eggs, except human beings.

He first met Charles Darwin on the latter's return from South America in 1836. The writer Thomas Carlyle asked to be introduced to Owen in 1842. He visited Sir Robert Peel and enjoyed the support of Gladstone in his plans for the new Natural History Museum. In the arts, he knew the painter JMW Turner, the poet Lord Tennyson and the writer Charles Dickens. Dickens and Owen actually exchanged signed copies of their respective works *Our Mutual Friend* of 1864/5 and *Memoir on the Gorilla* of 1865 for, in the former, Dickens had created the character of Mrs Podsnap specifically with Owen in mind. In Chapter II of the book he describes how:

> 'The great-looking glass above the sideboard … Reflects Mrs Podsnap; fine woman for Professor Owen, quantity of bone, neck and nostrils like a rocking horse, hard features, majestic headdress in which Podsnap has hung golden offerings.'

Other influential friends included Lord John Russell, the writer George Eliot (Mary Ann Evans) and the artist and critic John Ruskin. Owen also advised the missionary and explorer Doctor Livingstone on the writing of *Missionary Travels and Researches in South Africa* and met General Gordon on a visit to Egypt in 1874.

There are few clues to Owen's future eminence in his earlier years in Lancaster. Indeed, he was regarded by one of his tutors at the grammar school as lazy and impudent and likely to come to a bad end. He rose, though, to become one of the 'first six boys' at the school, which entitled him to claim a fee as an attendant at weddings in the Priory Church. This could be quite lucrative with fees ranging anywhere from a shilling to a couple of guineas.

After his largely undistinguished period at the grammar school, Richard Owen junior was apprenticed in 1820 at the age of 16 to Leonard Dickson, a surgeon and apothecary, to himself learn the skills of a surgeon apothecary and man midwife. Dickson, though, died in 1822 whereupon Owen was transferred to Joseph Seed. This period lasted for only a year until Seed became a naval surgeon. Finally, Owen joined James Stockdale Harrison whose work included carrying out post-mortems in the county gaol housed in Lancaster Castle.

His time with Harrison clearly left an impression, since it was the source of at least two tales passed on within the Owen family. His experiences were centred on the prison hospital housed on the top floor of what Owen called Hadrian's Tower.

The first tale concerns an evening visit Richard Owen had to make to one of the patients in the castle to administer some medicine. There had been several cases of fever, and that morning Owen had attended the dissection of a man he had treated earlier but who had died. He found this a difficult procedure to endure since he regarded the dissection as a kind of desecration of the dead man.

His evening visit was at 9 o'clock in late November. A storm was rising and the full moon could be seen between scudding clouds. At the gate, Owen declined the turnkey's offer of company in case 'the young doctor' was to be thought of as being afraid. At the foot of the stairs in the tower the wind came whistling down from the slit windows higher up. The whole tower was acting like a great organ pipe and Owen decided to try whistling himself to keep his spirits up. Into his mind, though, came visions of that morning's dissection – the pale collapsed features and the half-opened glassy eyes of the man. Into his mind too came thoughts of ghosts shimmering before him in the half-light of the stairwell.

In the end, though, it was Owen's rational mind that won the day. The ghosts were no more than sheets hung up to dry, animated by the draught and lit by the sporadic moonlight. Add to this the events of the day and the tricks of the mind and you have all the ingredients for a classic ghost story. The effect of witnessing the dissection did, though temporarily, make Owen resolve to give up a profession that could only be learned through such practices.

Richard Owen's second tale from this time arose only six weeks later so he had clearly overcome his previous reservations. A negro sailor who had been involved in the West Indies trade suffered an apoplectic fit following a fall during a drunken brawl. He subsequently died in the prison hospital. Owen had by that time begun putting together his own anatomical collection. He already had a human skull as well as cat and dog skulls and skeletons of mice and other small animals. Owen had also been reading a book called *Varieties of the Human Race* and decided that he would like to add the negro's head to his collection.

This would involve returning to the castle to remove the head, hiding it on the way out and also, by means of a bribe, gaining the co-operation of the turnkey so that a blind eye could be turned to the state of the corpse before the coffin lid was finally screwed down. All went well on a frosty evening and Owen descended the cobbled entrance to the castle gates for home, his prize safely wrapped in a brown paper bag.

The scene changes to one of the old cottages that nestled close to the castle walls. By the light of the fire sat two women: one the widow of the sailor, the other his

THE UNDERWORLD
The Underworld is full of ghouls, warlocks and witches – it's like Disneyland.

daughter and they were discussing the slave trade. Their conversation was interrupted by a commotion outside as the door was knocked open by a heavy blow. An object of horror then entered the room, coming to rest in front of the fire, its eyes glinting in the firelight. The mother fled screaming into the adjoining bedroom while the daughter looked down to see the face of her dead father. The drama ended as a black cloaked figure appeared, snatched up its head and disappeared back into the night.

What had happened, of course, was that Richard Owen had slipped on the icy cobbles and dropped his precious parcel, which had then rolled, down the hill. Owen recovered himself and rushed off in pursuit.

Though a somewhat lurid tale which may have gained much in the telling and re-telling within the Owen family (the above version is largely as recalled by Owen's grandson) it does have a certain truth about it. Owen later became a prolific anatomist, conducting large numbers of studies using material drawn from a variety of sources. These included dissections of animals that had died in the gardens of the Zoological Society. His study of medicine at the University of Edinburgh from 1824 also fell shortly before the infamous case of Burke and Hare who in 1827/8 had committed murder in order to supply cadavers for dissection at the medical school. The whole study of human anatomy was gaining a macabre air in the popular imagination of the day.

At Edinburgh, which was a national centre of excellence in medicine, Owen studied anatomy under Dr John Barclay as well as courses in the practice of medicine and midwifery. Owen transferred in 1825 to St Bartholomew's Hospital at the suggestion of Barclay and passed the examination of the Royal College of Surgeons in 1826. He then set up in private practice in London.

When Owen returned to Lancaster in 1842 for the public dinner given in his and William Whewell's honour, he too claimed the credit for securing a day off for the boys at the grammar school. Owen recalled in his speech that evening how the granting of such holidays had made an impression upon him and how in his day they had marked the victories of Wellington and the successes of his friend Whewell at Cambridge.

Richard Owen was introduced to the gathering in the Assembly Rooms as the Hunterian Professor and Curator of the museum of the College of Surgeons. He was appointed as the first Hunterian Professor in 1836, and in 1842 (the year of his visit to Lancaster) became joint conservator of the museum alongside William Clift.

Owen's impressions of the evening are recorded in a letter to his wife written on 7th September.

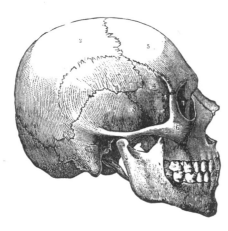

> 'I take up a happy pen this morning to tell you that the dinner concluded to the highest satisfaction of everyone who partook of it… As we walked in procession to the Town Hall, Mr Whewell and the Mayor, then the MP for the town and myself, and the rest two and two, we were cheered on by all the humbler folks, and when we sat down to a most princely banquet – the chairman, Whewell, and I, on three raised state-seats at the head – we were greeted and complimented in a truly English and manly manner by the ablest men to whom the proposing of the toasts had been assigned.'

Richard Owen was a regular visitor to the Lancaster area up to the death of his mother Catherine in 1838. His most lasting contribution to the life of Lancaster arose from his period as a member of the Commission of Inquiry into the Health of Towns from 1843–46. In 1844, he wrote a report about Lancaster's sanitation, which noted that there was neither a public nor private water supply in the town and that the wells and river on which people relied for their water were both polluted. Owen also reported how the town's sewers, which mostly drained into an old mill race, were still open. These sewers could also cause flooding if they became blocked. Lancaster was therefore very vulnerable to epidemics such as cholera which could affect up to 10 per cent of the population. The result of the report was the provision of a safe water supply for Lancaster with its first reservoir being completed in 1855.

Edward Frankland

Professor Sir Edward Frankland held a series of chairs of chemistry including the first at the newly founded Owens College in Manchester, the Royal Institution and the Royal College of Chemistry. He was primarily an organic chemist but is remembered as having introduced the concept of valency into chemistry (the numerical rules governing the combination of atoms into molecules). His notation, representing molecules and their structure, is furthermore still in use today. Like William Whewell and Richard Owen, Frankland can also be regarded as a wordsmith since he is sometimes credited with naming the element helium discovered in the solar spectrum by Norman Lockyer, with whom Frankland worked.

In 1859, Edward Frankland and the mathematician John Tyndall became the first people to spend a night on the summit of Mont Blanc. Their observations of the burning of candles there led Frankland to investigate the influence of atmospheric pressure on combustion. In the same year, Frankland began a long study of the problem of water purification when he and Professor August Wilhelm Hofmann were asked to report on means of deodorising sewage. Raw sewage was then released directly into the River Thames, which was described as 'black and horribly offensive.' Frankland returned to water purification in 1865 considering various issues, including water analysis, river pollution and domestic water supply. His work on the filtration of sewage and bacteriological processes was of fundamental importance. It was for his services as a water analyst that Frankland was knighted in 1897 on the occasion of the diamond jubilee of Queen Victoria.

This distinguished career is all the more remarkable given the circumstances of Edward Frankland's early life. He was born illegitimate, was brought up by his mother alone for nearly six years and had by the time he was eight years old attended no fewer than six different schools.

In Lancaster, Edward Frankland was sent to a private school run by James Willasey where he was taught English, arithmetic, writing, geography and history as well as being given lessons in French privately. The school contrasted with others in that it did not concentrate on classical and religious studies.

WATER

Water keeps everything fresh – flowers, blood for mosquitoes, everything.

EXCRETION –
VOIDING THE BOWELS – POO

When you grow up, the invisible trap-door in your bottom falls off. Poo is toxic, it is made of mouldy food. Does all animals' poo make you go blind like cats' and dogs' poo? Toxicara makes you go blind. Toc-sick-cara. Toxic is poisonous – sick like flu, and cara because it makes you go blind. That's what the word means, and why it is the word.

Edward Frankland

Willasey encouraged Frankland's growing interest in science. At the age of 10, Frankland had borrowed a copy of Priestley's *History of Electricity* from Lancaster Mechanics Institute. Inspired by this he made an electrophorus, which he used to charge his home-made Leyden jar. He also made an electric machine and a voltaic pile. He began experiments in chemistry, generating hydrogen by the action of sulphuric acid on zinc. He made fireworks and did experiments in a room provided by the parents of his friend Richard Turnbull.

In 1837, Frankland moved on to the grammar school. The main subject was Latin and most of those attending aimed to advance themselves by moving on to university, the church or into law. The teaching was generally mechanical and uninteresting, and the year's prize books were allotted not for merit but on the casting of a die. The regime was relieved by weddings at the Priory at which sixth form boys acted as attendants, and public hangings held across the castle green before the start of the school day. Frankland himself witnessed between 15 and 20 hangings, including one of a woman and a triple hanging.

In January 1840 at the age of 15, Edward Frankland began a six-year apprenticeship to one of Lancaster's four druggists, Stephen Ross. Ross was chosen for what was seen as a cheap back-door route into a medical career. He was also chosen for his piety as a *'burning and shining light'* in the lowest section of the Church of England though, for Frankland, the main advantage of Ross' piety was that the shop didn't open on a Sunday. Writing about the experience some 50 years later, he looked back over his apprenticeship as *'six years' continuous hard labour, from which I derived no advantage whatever, except the facility of tying up parcels neatly'*.

That long and wearisome six years' apprenticeship began each day at 5.45 when Frankland collected the keys, took down the shutters, swept the floor and dusted the bottles in preparation for Ross' arrival at 8am. The routine varied in the winter with a later start at 6.45am but the additional need for the fire to be lit. The shop closed at 8pm. With half an hour allowed for breakfast and an hour for dinner, Frankland worked a 77-hour week in the summer and a 71-hour week in the winter.

His early duties were indeed menial. He had to collect and roll barrels of treacle along the streets and carry two hundredweight (51 kg) sacks of barley up the shop's steep and narrow stairs. He had to use a 20-pound (9 kg) mortar and pestle for grinding. This also involved wearing a linen sack over his head as a mask against the dust from the drugs, paint, sugar and cocoa, which had all to be ground. This process was often slow and laborious with a day, for instance, being needed to grind a pound (0.5 kg) of cocoa ready for sale. Frankland would also mix paints and blend sugar after grinding. The most difficult task with the mortar and pestle was the preparation of mercury ointment. This involved mixing 14 pounds (6 kg) of hog's lard with six pounds (3 kg) of mercury so intimately that no globules of mercury were visible with a magnifying glass. The process could take as long as three months with mixing going on at a rate of two days per week. In the shop, Frankland would sell tea, sugar, coffee and nutmeg as well as the three ingredients for shoe blacking – bone black, treacle and oil of vitriol (sulphuric acid).

After two years, Frankland progressed to wearing a white apron rather than a black one, had his hours reduced by half an hour a day and was granted one week's

The Leyden Jar.

THE EARTH'S MANTLE
We live on a crust – this whole house, the
garden, are all on a crust of the Earth.
Below us are big, dark caverns and lava
and volcanoes in the centre of the Earth.
You have to go down deep, dark caverns
to get to the centre of the Earth. No people
go there because they'll be burned to bits
and they'll be as dead as a dingbat.

annual holiday. His duties now included polishing the counter, seeing to the window display, painting the shop, serving medicines and his 'chief delight', pulling teeth. Frankland would also unofficially prescribe medicines, his customers often having more confidence in him than in their own doctors.

While working at Ross' shop on Cheapside, Frankland was introduced to Christopher Johnson who was the surgeon at the dispensary. His sons Christopher and James were also doctors and shared a practice on St Nicholas Street round the corner from Cheapside. The Johnsons took an interest in the welfare of apprentices by arranging lectures on scientific subjects and lending books through the Mechanics Institute. In about 1840, James Johnson made a house available for use as a chemical laboratory and lecture room. Christopher Johnson senior taught Frankland chemistry and did much to encourage his interest.

In October 1845, Frankland was released from his apprenticeship and transferred to the Museum of Practical Geology in London to study under Dr Lyon Playfair, having been introduced to Playfair by Christopher Johnson. Another Lancaster connection is that the town's MP, Thomas Greene had his London address two doors away from Playfair's laboratory.

On 23rd October 1891, Edward Frankland visited Lancaster for the opening of the Storey Institute. Frankland recalled the influence on him in Lancaster of James Willasey and Christopher Johnson and noted that whatever knowledge he possessed up to the age of 21 he owed to Lancaster. He particularly picked out the Johnsons for praise, for, 'they not only lent me and other youths valuable chemical apparatus, but they were ever ready to give advice and to smooth away difficulties.'

William Turner

A plaque in the wall of 7 Friar Street, Lancaster announces:

> 'In this house was born on the 7th day of January 1832 Sir William Turner KCB FRS DSc LLD DCL Knight of the Royal Prussian Order pour le merite Professor in and Principal and Vice-Chancellor of Edinburgh University.'

Professor Sir William Turner was associated with Edinburgh University for some 62 years beginning as a demonstrator (effectively a private assistant to Professor John Goodsir) but rising to first succeed Goodsir in the chair of anatomy and then become principal of the university itself. Indeed, he was the first Englishman to ever become head of a Scottish university. At his death, Turner's achievements in anatomy were rather generously compared with those of the great French zoologist and anatomist Baron Georges Cuvier who had died in the year of Turner's birth. Cuvier had learned how to reconstruct extinct animals from just a few of their bones and had applied the method of classifying living animals to fossil remains of mammals and reptiles. However, Turner's scientific talents lay more in accurate measurement and careful recording than such bold leaps of the imagination.

Turner was also a successful administrator, doing much to expand Edinburgh University physically and academically. He represented the Universities of Edinburgh and Aberdeen on the General Medical Council and became its president, the first

Scottish representative to do so. He also served on a royal commission to consider reform of the conditions required for entry into the medical profession. Turner argued for the superiority of the universities over a proposed central examining body and, though his was a minority view, it prevailed when the findings of the commission were formulated into law.

Turner's father, also called William, died when young William was five years old. William started his education at a preparatory school for the grammar school. However, by the time he could have transferred there, the grammar school's fortunes had so declined that it was no longer an obvious choice. It was described as being, 'not in a very flourishing condition, either as regarded its general administration or in its situation.' Instead, at the age of 10, William became one of 15 boarders at a private school run by the Reverend William Shepherd at Long Marton, near Appleby.

William's return to Lancaster at the age of 15 saw him apprenticed to his uncle John Aldren, who was a chemist. Within a few weeks, though, Dr James Johnson arranged for Turner to switch his apprenticeship to his brother, Christopher who, like his father, was a surgeon at the dispensary. Turner's duties were to receive messages, make appointments and assist in the keeping of the accounts. He would make up pills

BUBBLES

If there was air in the bubbles but no air in the sky – when you blew the bubbles, it wouldn't pop because no hole would be made in it.

William Turner

and roll them (they would have numbered thousands), make up bottles of medicine, dispense drugs and make up prescriptions. He also made poultices, applied leeches for the bleeding of patients, learned to apply bandages and splints and carried out minor surgery. He recalled how on one occasion he had to carry a wrapped amputated limb in a basket to the Johnsons' surgery one Sunday morning. As he passed people leaving church he imagined they could see through the wrappings and were aware of the gruesome object he was carrying.

Turner regarded both William Whewell and Richard Owen as examples to live up to, and dedicated his *Memoirs upon Whales and Seals* of 1888 to Owen. Owen mentioned in his acknowledgement that he had not returned to Lancaster since the death of his immediate relatives.

William Turner completed the final year of his five-year apprenticeship at St Bartholomew's Hospital in London. In this, he was following in the footsteps of Owen, and he obtained his diploma in anatomy, physiology and chemistry in 1853.

At the opening of the Storey Institute in Lancaster in 1891, Professor Sir William Turner was welcomed by Sir Thomas Storey. In seconding the vote of thanks to Sir Thomas, Turner spoke of his early days in Lancaster and eloquently recalled in the presence of Christopher Johnson junior that,

> 'There was in those old days … a small lamp of science hung in Lancaster, and those who lit it were the family of Johnsons.'

There is no single reason why Lancaster should have produced so many scientists of such significance during the first half of the 19th century. Their backgrounds and routes of progression were quite varied yet all were probably inspired or encouraged by some individual personality. Whewell was in a sense discovered by the Rev Joseph Rowley. Owen, despite a lacklustre schooling and fitful start to his apprenticeship, clearly underwent some kind of transformation while apprenticed to James Stockdale Harrison. As members of an earlier generation, Whewell and Owen also served as role models for Frankland and Turner as they rose through school and apprenticeship. Both Frankland and Turner were considerably assisted by the Johnsons, and James Willasey was a further influence on Frankland. Even Stephen Ross, despite Frankland's scathing comments in later years, provided a basic grounding for Frankland's work in the laboratory.

Public and popular

As Whewell and the others were finding their way to careers in science, Lancaster itself was catering increasingly for those with a more general appetite for the subject through popular public lectures, learned societies and, towards the end of the 19th century, a public astronomical observatory. While this activity inevitably lagged behind that of London and even other regional centres, it did follow their pattern of development. A trend is apparent in Lancaster towards more specialised interest groups, especially in the field of natural history, from about 1870 onwards. At the same time, societies were aiming to widen their membership through lower subscriptions, and this trend continued into the 20th century.

WATER TENSION
You can pop water (droplets on grass). You touch it and it sort of dissolves onto your finger. When you touch them (droplets) they disappear and form a little droplet on my finger.

Lancaster's first society catering in part at least for those with an interest in science was its Literary, Scientific and Natural History Society of 1835. With a subscription of one guinea, it was aimed at a wealthier clientele, but still managed to attract between 60 and 100 members.

After 1848, the activities of the Literary, Scientific and Natural History Society were absorbed into the programme of the Lancaster Athenaeum whose broad aims were to entertain and instruct the public, the latter role being fulfilled through lectures on literary and scientific subjects.

A frequent visitor to the Athenaeum was the Rev St Vincent Beechy. His first pair of lectures delivered in 1849 proved so popular that some members of the audience could only be accommodated behind the projection screen. Beechy's mode of presentation created as much interest as his subject, in particular his Trinoptric Lantern, which could project three images simultaneously and allowed images to be dissolved into one another.

A rather different experience was provided at the Athenaeum in 1850, by JP Childe, in the form of a 'Grand Scientific Entertainment – when the results of Astronomical Research will be communicated in a popular Exposition of the Solar and Stellar Creation'. Childe did this by employing such devices as a planetarium comprising 30 scenes each of 324 square feet, a panorama of the heavens exhibited with the Leviathan hydro-oxygen microscope – magnifying seven millions of times – and a cyclorama.

His presentation seems to have lived up to its promise. A reporter recorded that the hall 'was crowded to excess, and the lecture was lucid and instructive, if we except some occasional murderings of the Queen's English'. In some respects the event was too popular, for, 'some parties were unprincipled enough to fire off crackers, and otherwise annoy the audience during the time the hall was in darkness' (the entertainment was a few days after Bonfire Night).

1885 saw Sir Robert Ball (the Patrick Moore of his day) present *An Evening with the Modern Telescope* to a capacity audience in Lancaster's Palatine Hall, in a lecture arranged by Lancaster Mechanics Institute. Another five such lectures followed from 1891–1899.

This period also saw the formation of Lancaster Philosophical Society. With a subscription of five shillings, this was more generally affordable than the Literary, Scientific and Natural History Society had been. A lecture in 1890 by the Rev TE Espin (the first Director of the British Astronomical Association's Spectroscopic and Photographic Section) anticipated the opening of Lancaster's Greg Observatory and outlined the contribution of amateurs to astronomy.

The Greg Observatory

Lancaster's public astronomical observatory, the Greg Observatory, was opened amid great ceremony on 27th July 1892.

Yet today, only the building's foundations mark the site, largely unexplained, in an obscure corner of Williamson Park. So, how was it that Lancaster came to have such an observatory in the first place and, perhaps more significantly, how did it come to lose it again?

The Gregs

In common with other members of the Greg family, John Greg, manager of Low Mill at Caton near Lancaster, had an interest in astronomy. His brother, Robert Hyde Greg, built an observatory at Norcliffe Hall while his nephew, Robert Philips Greg (son of Robert Hyde), retired from the family business in 1871 and devoted much of his time to astronomy. He was particularly interested in meteors and meteorites, and is credited, along with Professor Alexander Herschel, with the discovery of the Geminid meteor shower, which can be seen on, or about 14th December each year. Another of Robert Hyde Greg's sons, Edward Hyde Greg, also had an observatory, this time at Quarry Bank.

John Greg set up a wooden observatory in the grounds of his home, Escowbeck House. This housed a seven-inch (18 cm) refracting telescope made by Thomas Cooke and Sons of York, mounted atop a massive iron pillar, which was in turn set upon a solid stone foundation. The telescope was equipped with a range of eyepieces to give a selection of magnifications, and a Maclean spectroscope (a device to split light into its component colours). The observatory also had a smaller transit telescope set on a massive stone pedestal, and a sidereal clock to tell the time by the stars.

John Greg died on 23rd November 1882, aged 80, and was buried in the churchyard at Brookhouse. He had been a member of Lancaster's first scientific society, Lancaster Literary, Scientific and Natural History Society, from 1836. Shortly after joining, he presented the society's museum (housed in what is now the Grand Theatre) with a hundred catalogued specimens of Scottish rock and several stuffed birds including two birds of paradise, an osprey and a whiteheaded eagle. Later he gave the museum 'a tame swan'. Greg was also a member and patron of Lancaster Athenaeum, which included scientific lectures in its programme. He presented a gold medal for mathematics at Lancaster Royal Grammar School and was a Manager for Life of Lancaster Mechanics Institute.

The Park, Lancaster

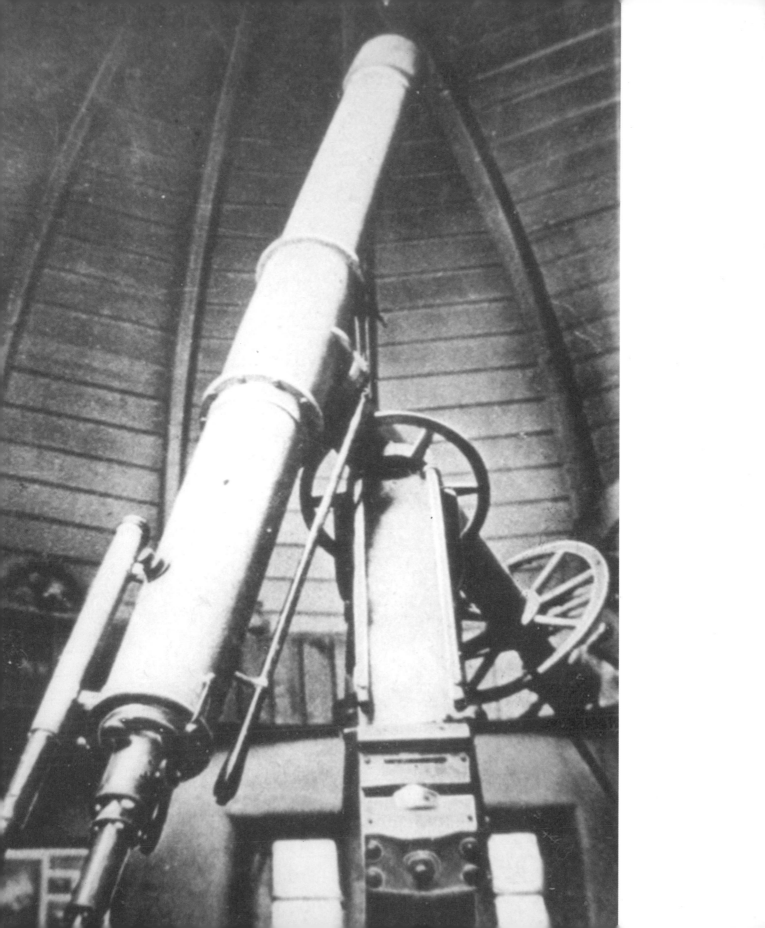

The Observatory

Following his father's death, Albert Greg offered the telescope as a gift to Lancaster Corporation. Such a gesture was quite common at the time since the market for large and expensive instruments was limited. Many towns set up public observatories to house such gifts, and observatories became to an extent an expression of civic pride.

However, Lancaster cannot be said to have hurried to take up Albert Greg's offer. By late 1889 when a site for the observatory was being sought, the *Lancaster Guardian* commented that the process had already become drawn out. A site in Williamson Park was chosen. A visit to the old observatory at Escowbeck, meanwhile, by a Mr Normanton from Thomas Cooke and Sons, of York resulted in the suggestion of combining the observatory with a meteorological station, which might be of more practical benefit.

It was also proposed that rather than transfer the telescope directly from Caton to Lancaster, it might first be returned to York for cleaning. This would cost 29 pounds 10 shillings. For a further 15 shillings per day plus expenses, a representative from Cooke's could attend and make any necessary adjustments to the instruments.

By the summer of 1891, a plan for the observatory building had been prepared by the Borough Surveyor, John Cook. A glazed outer room lay at the western end. Here, small telescopes and binoculars could be used to view the local scenery from the Fylde coast to the hills of the Lake District. The room also came to house a small collection of curios including a Zulu spear and shield, First World War gas masks and a stuffed snake. The central room housed the main telescope beneath its original dome. A room at the eastern end of the building housed the transit telescope and sidereal clock.

A north-south line (vital for the proper setting up of the telescope) was struck off by the Rev John Bone, a local amateur astronomer, and Mr Cook on 4th September 1891 and marked by a chimney pot on one of the cottages in the nearby area of Golgotha. The Rev Bone became the observatory's first Honorary Director and gave or procured various maps, charts and catalogues for its library.

In his opening address, Dr Ralph Copeland commented on the importance of public observatories both for education and for the advancement of science. He noted that the work of the professional observatories of the time was of an exacting but routine nature. He felt that there were many problems that could be at least partially solved by dedicated amateur astronomers working with equipment such as that housed in the Greg Observatory. Dr Copeland was presented with a silver key to the observatory by Councillor Molyneux, Chairman of the Observatory Committee. This had been made by the Lancaster jeweller, Atkinson's and bore on one side Lancaster's coat of arms and on the other a commemorative inscription.

Dr Copeland's themes were expanded upon in an address given by the Rev John Bone to a meeting of The Lancaster Philosophical Society. The Rev Bone was greatly impressed by the interest shown in the heavens by the working people of the town and felt that by seeking greater knowledge they were looking beyond the limited horizons of their material existence. Also, such observatories might lead people eventually to pursue a career in astronomy. The Rev Bone spoke too of the use professional astronomers might make of accurate series of observations made by amateurs.

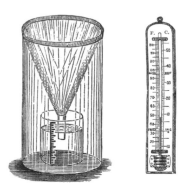

PAPER WEATHER VANE

A strip of paper acts like a weather vane –
as the wind moves, the paper moves...

The first curator of the Greg Observatory was George Ingall. He took up his duties on 1st May 1892 (before the official opening) but was not paid a salary. Instead, he was allowed to charge visitors to the observatory one penny. Ingall also began taking weather readings over an experimental three-month period though the weather station proper (a third order weather station consisting of a limited number of instruments to be read once a day) didn't begin operation until 1st January 1895.

The weather station comprised a barometer, maximum and minimum thermometers, wet and dry-bulb thermometers, a sunshine recorder, pluviograph (an instrument measuring the amount of water to be accommodated by the drains) and a self-recording rain gauge. An anemometer was sited nearby. Some of the instruments were kept within a railed enclosure near the observatory. Readings were taken daily at 8.30am and weather reports displayed at the observatory itself and the Storey Institute. When the new town hall opened on Dalton Square, the latter report was posted on a notice board on the side of the old town hall buildings on New Street.

Various names and initials can be seen accompanying the weather records. Among these are William French (or possibly Finch), K. Rees, TC Joyce, WH, GAW and JCT. April 1903 sees the first appearance of the initials JD, James Dowbiggin, who was to be associated with the observatory for the majority of its existence.

The instruments in the railed enclosure were particularly vulnerable to theft or to people disturbing or breaking them, and the weather records report many such instances. Some of the earliest such reports were of the anemometer being 'found unworkable' in 1912 and the grass minimum thermometer being broken one evening in August 1914. The weather station underwent an annual inspection by the Air Ministry that had, during the First World War, absorbed the Meteorological Office. The Air Ministry inspector had to check that the readings were being properly taken, that – as far as possible – the record was continuous and that anything that might affect the readings (such as overshadowing vegetation) was being controlled. In November 1939 the weather station failed its Air Ministry inspection.

In July 1905 the post of observatory curator was advertised as being vacant. George Ingall is known to have tendered his resignation in April 1902, so perhaps no one else was prepared to take on the job for the penny admission charges alone. The advertisement includes mention of a weekly salary of 10 shillings and the hours of attendance at the observatory as 11am to 12.30pm and 1.30pm to 5pm in the summer and 11am to 12.30pm and 1.30pm to 3pm in the winter. Of the three applicants, the successful candidate was James Dowbiggin. Dowbiggin was a particularly appropriate choice because he is known to have made frequent use of the telescopes as a young man in John Greg's original observatory at Escowbeck.

It was possible to use the telescopes in several ways. Casual daytime visitors could look through the main telescope at the planet Venus or perhaps a bright star. More organised tours of the heavens were led by the Rev Bone or Mr Tom Taylor of Market Street who would use the telescope to illustrate talks given to groups of visitors. People could also use the telescope individually under supervision and, finally, suitably qualified individuals could use the telescope unsupervised for their own private study. More than a hundred people made use of the telescope in this last way. Mr Taylor used the telescope 158 times and others seem also to have made regular use of it, suggesting the existence of a group of active amateur astronomers in the town up to the 1930s.

James Dowbiggin was best known in Lancaster for his job of posting the daily weather reports for which he gained the nickname *Weather*. The astronomical role of the observatory is not so well recognised or documented.

Up until his death in 1906, the Rev Bone was the observatory's main astronomical influence. He was a Fellow of the Royal Astronomical Society, ran classes in astronomy at the Storey Institute and was an important figure in both Lancaster Philosophical Society and Lancaster Astronomical and Scientific Association. He was succeeded as Honorary Director of the observatory by Lancaster's Coroner, Neville Holden.

The one major astronomical event to have left its mark was the total eclipse of the Sun of 29th June 1927. The track of totality crossed north Wales and northern England with the centre line from which totality was of longest duration (23 seconds), passing over Southport and Hartlepool. Lancaster lay at the northern edge of the track of totality. A report of the day's events was prepared for the Greg Observatory Committee:

> 'Observations taken in 1926 and confirmed this year showed that the Memorial Building would be in a direct line between the Sun and the telescope during totality and this prevented any use being made of the Observatory. A good point of view was found in the extreme SE of the Park overlooking the Quernmore Valley and here about 80 people assembled just before the time of 1st contact at 5.20 ST.

Observations made by Mr J Dowbiggin with a powerful field glass showed that the first contact was dimly seen but the time could not be accurately determined owing to a haze or cloud. Conditions improved during the first part of the eclipse but just before the time of totality a strata cloud passed in front of the Sun's disc and prevented both the shadow and Corona being seen. The duration of totality was estimated at about eight seconds. After totality conditions improved and the later phases of the eclipse were fairly well seen.'

Prior to the eclipse, James Dowbiggin gave lectures about what might be expected, such as that given to a well-attended meeting of Morecambe Astronomical and Astrological Society.

Partial eclipses of the Sun were also followed from the observatory on 30th August 1905 and 17th April 1912.

Lancaster Royal Grammar School

The year 1939 was an important one for the observatory. It saw the weather station failing its Air Ministry inspection. It also saw the death of its Honorary Director, Neville Holden but, most important of all, it saw the retirement of its curator, James Dowbiggin. Dowbiggin was by then 80 years old and suffering increasingly from ill health.

Lancaster Corporation chose not to replace Dowbiggin but rather to reach an arrangement with Lancaster Royal Grammar School. Under this, the school would make use of the observatory and continue to take weather readings. In the event of any weather instruments being lost or broken however, these would not be replaced. This meant that by 1943 only the rain gauge remained. Members of the public would still be able to visit the observatory by arrangement. The Corporation in turn undertook to maintain the fabric of the building while the sidereal clock, which had been affected by the damp conditions in the observatory, was to be removed to the school.

Lancaster Royal Grammar School Astronomical Society held its first meeting in the Greg Observatory on 6th June 1939. The outbreak of the Second World War meant that the younger masters at the school left to serve in the armed forces, so the running of the astronomical society seems to have been left largely in the hands of the senior boys. Nonetheless, their book of observatory recordings begun in 1941 shows that they made regular use of the telescope. Solar observations were made on over 60 occasions and the observatory was used on more than 50 nights. Most of the observations were of the moon, planets and double stars, though several eclipses, the planet Uranus, the minor planet Vesta, disappearances of stars behind the moon, and Comet Whipple, seen in the spring of 1943, were also recorded.

Use of the observatory was not without its difficulties. Damp had caused some of the woodwork to rot, especially round the base of the dome. This was a particular problem because the wheels of the dome would often come off the metal rail on which they ran and it was difficult to find sound wood to which to fix the rail. The canvas covering of the dome had also deteriorated, allowing rainwater to drip into the building and cause condensation on the telescope lens.

Another difficulty was ongoing vandalism. Access could be gained to the observatory through the slit in the transit room roof, which was particularly difficult to

WIND MACHINE
You need a feather, a wind bell and some wood. The feather has to be put on a hook at the top. The wind bell has springs. When you don't need it anymore, you break off the feathers and it doesn't work anymore.

secure. At one point, the schoolboys mounted guard at the observatory in an attempt to catch the culprits but this proved to be only a temporary solution.

The main difficulty was with the mechanism of the telescope mounting. Locking devices broke and the telescope became increasingly difficult to manoeuvre. Nonetheless, the boys continued using the telescope in this condition for some 18 months, which probably aggravated its condition. They seemed reluctant to report the problem, concentrating instead on the superficial tasks of keeping the observatory tidy. When they did report the mechanical problems, they hesitated over whether to blame it on the damp or intruders.

After many delays, a letter was prepared for the City Engineer in late 1943 attributing the damage to the effects of damp. After receiving no reply and having sent a second letter, a Mr Hannah from Sir Howard Grubb, Parsons and Co Ltd (who had taken over Cooke's business in York) finally inspected the instruments during the summer of 1944. A letter dated 20th July stated 'the instruments are in a terrible state of repair'. The main question that arose was – are the instruments essential for carrying on the war or, in other words, are regular meteorological records requested by headquarters? If not he could not recommend his firm to undertake the overhaul until after the war.

In the event of the instruments being repaired it would mean a technical adviser coming down from Newcastle to make a minute inspection and later, the instruments being sent to Newcastle. As a rough idea of the price he said:

> 'TWO HUNDRED POUNDS WOULD NOT COVER THE COST OF REPAIRING THE 8-INCH TELESCOPE.
>
> Regarding the solar transit instrument – this is nothing now but a brass tube without either eyepieces or objective lens, and the stand and fine adjustment is also fouled.
>
> We searched the premises to find the spectroscope but failed to do so.'

In August 1944, the headmaster, Mr Timberlake, ordered the Astronomical Society to stop using the observatory and, on 19th September, it was closed for the duration of the war. From then on, the observatory seems to have been left largely abandoned. Vegetation began to grow from the surrounding slopes, eventually hiding it from view, while the effects of the weather and continuing vandalism took their toll.

Eventually, in the late 1950s, tenders were sought for the demolition of the observatory along with the park bandstand.

Closing

The closure of the Greg Observatory was no doubt a source of embarrassment to the grammar school and, to a degree, to the Corporation of Lancaster, although the special conditions of wartime and the period following meant that priorities lay elsewhere. Today, the foundations of the observatory and the overgrown remains of the weather station stand as puzzling memorials to all who contributed to their creation and decline.

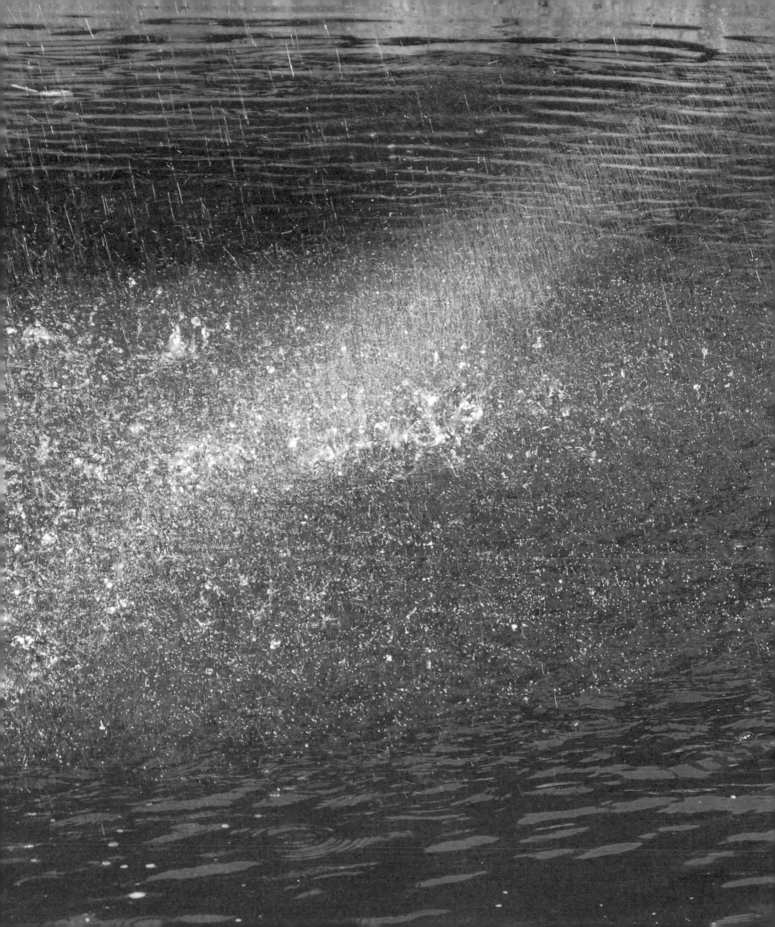

Chapter 4. THE PARK AS ARTEFACT: ART, ARCHAEOLOGY AND INTERPRETATION

David A Barrowclough

Introduction

Chinese porcelain, French furniture, Japanese calligraphy – the drive to collect seems to be a near universal human trait, and nowhere is it more rarefied than in the world of art where connoisseurship finds its apogee. Whether it be Renaissance sketches, African tribal art or Polynesian bark paintings, there are experts ready to evaluate, advise and to value. As an archaeologist I caught the collecting bug at an early age (for a discussion of the link between collecting and archaeology see Schnapp, 1996:11) and I am therefore naturally drawn to those contemporary artists whose practice has been to question the nature of collection, curation and display and thereby subvert commonly held notions of expertise. Amongst these artists are those who share my interest in archaeology, notably Susan Hiller, Mark Dion and particularly Robert Williams.

In this chapter I will consider the relationship between the collector and the collected as informed by *Thesaurus Scienta Lancastriae*. I will deconstruct the collecting process, identifying its constituent parts, and question why it is that archaeologists collect certain objects or artefacts and not others, distinguishing between cultural and natural objects. I argue that the artists of the present work in fulfilling their role as social critic have subverted the traditional distinction between Nature and Culture. In so doing they have freed archaeologists from the constraints of the established paradigm, and opened up an opportunity to approach the past in alternative ways, from which we can gain fresh insights into past worlds. Using an archaeological case study to illustrate my argument I show how there are clear parallels between the artistic practice of Jack and Robert, and contemporary archaeological practice. Breaking down boundaries between natural and cultural objects allowed the Williamses to gain a richer understanding of Williamson Park than would have been possible had they restricted their collection to only one type of object. For archaeologists this points the way towards a new approach to fieldwork, which has come to be known as the 'life history' or 'object biography'. Here one attempts to understand the prehistoric site or object in a way analogous to that in which a biographer might gather information from numerous sources in order to reconstruct the life history of a person. I conclude that what archaeologists can learn from the *Thesaurus Scienta Lancastriae* is that in order to do this thoroughly one needs to break down the dichotomy between natural and cultural artefacts.

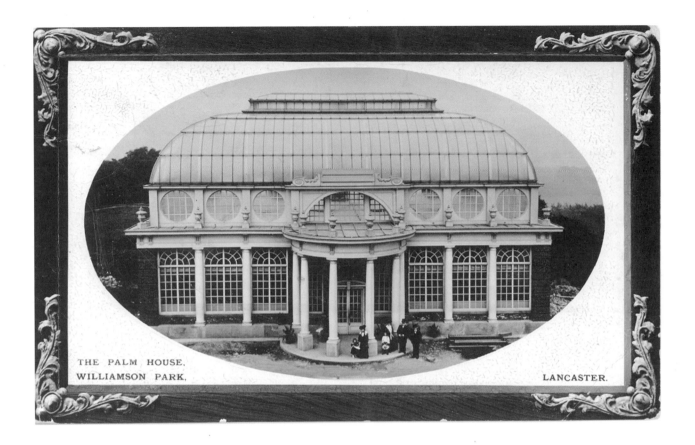

Artists as archaeologists: the Williams' practice

Robert Williams' most recent work is a collaborative piece undertaken with his young son, Jack. Over a period of a year they have devoted their holidays, weekends and evenings to the collection, curation and display of all manner of objects found within Williamson Park, in the City of Lancaster.

The labour has been apportioned between them, with Jack taking responsibility for both the selection and identification of artefacts, while Robert documented and displayed the collection.

What initially drew me to the piece were the clear parallels between what I do as an archaeologist and, what the Williamses have done in their art. In examining this it is necessary to deconstruct the collecting process into its four constituent parts:

- Discovery
- Curation
- Identification
- Display and presentation

Discovery

Both archaeologists and the Williamses work with found objects. I have been concerned with the ways in which one interprets those stray objects that farmers, metal detectorists, builders and general members of the public occasionally encounter in their daily routine and report to local museums (Barrowclough, in preparation). The discovery of these objects comes about in a similar manner to that in which Jack discovered the artefacts placed in the *Thesaurus*. A member of the public walking through a park, across a ploughed field or over a building site sees something in the corner of their eye that attracts their attention. They walk over, bend down and discover a hitherto unknown object.

Thus begins the process of collecting, first one object, then another and so on. Although individually these events are rare, over the last centuries the number of prehistoric artefacts found in this way in the north west of England is enormous (see Barrowclough, in preparation, for detailed catalogues of the objects found in this way). In the process of collecting, Jack, like every other collector, walked past and trampled over other artefacts, which he chose not to select for his piece. It may be, for example, that on one day he was only interested in collecting a particular class of artefact, as was the case on November the 6th when he collected fireworks.

On that day he chose *not* to collect other artefacts. In this way Jack's collection is a partial record of what was in any event a fragmentary resource, a situation familiar to archaeologists.

Curation

The second parallel in the practice of Robert and Jack's art that appeals to an archaeologist is that of curation. Robert has taken the objects collected by Jack, and he has cleaned and conserved them.

Then he has taken care to document them by means of a digital database, which records their physical dimensions, their size and weight, and also their appearance by means of photographs. All these are familiar to the archaeologist, for whom the careful handling and meticulous recording of objects are standard, but essential, procedures. It is the application of the scientific techniques of conservation and reconstruction, allied with methodical recording of the object, that removes the practice of archaeology from mere collecting, placing it in the realm of professional science and, thereby, accords it professional status.

Identification

Rather than follow the traditional approach, utilising a combination of published reference works and his existing knowledge to categorise them, Jack has identified and classified each of the objects by himself. He has grouped objects according to his own interpretation, placing like with like. Pot sherds with blue decoration have been grouped with other pot sherds with blue surface decoration, separate from those with text, which form a sub-group of 'Lancastrian Pottery' within the sherd typology.

This interest in taxonomy, which is at the heart of Jack's classification, has been a central concern to archaeologists since the beginning of the discipline in the 19th century. The work of John Evans in Britain, Flinders Petrie in Egypt and the German researcher Max Uhle in Peru did much to establish the chronology for these ancient civilisations based upon the classification of the archaeological finds according to type (Daniel, 1975).

It is a practice continued today in my own work (*ibid.*), which has involved the division of the various finds into first types, and then sub-types, in order to establish a typological sequence for the prehistoric material culture of the north west of England.

Display and presentation

The final parallel that impresses me is that of the display and presentation of the objects to the public. The presentation of their artefacts posed carefully so as to convey the meaning that the practitioner determined, bears a close relationship with the presentation of archaeological artefacts.

The garden shed of the Williams' piece successor to the *Wunderkammer* of the antiquarian collector. While the parallels with archaeology are clear, I will not dwell upon this issue in detail here as it has, in a different context, recently been the subject of detailed consideration by Colin Renfrew, to whom the reader is directed for further discussion (Renfrew, A. C. 2003, chapter 3).

The Williams' artistic endeavour therefore consists of a performance of collecting, or better, a series of performances which together constitute the processes of collection, the transitory nature of which leaves a permanent trace in the form of the objects, documents and display that result from the endeavour. The artistic project is not dissimilar to that of the archaeological, whose transitory intervention in the landscape is recorded for posterity in the form of logbooks, site reports, photographs, plans and, of course, the artefacts that they discover and display.

British Barrows.
a. Long barrow. *b.* Druid barrow. *c.* Bell-shaped. *d.* Conical. *e.* Twin barrow.

THE FIRST UNDERWORLD
The Earth-Gods made the first Underworld.

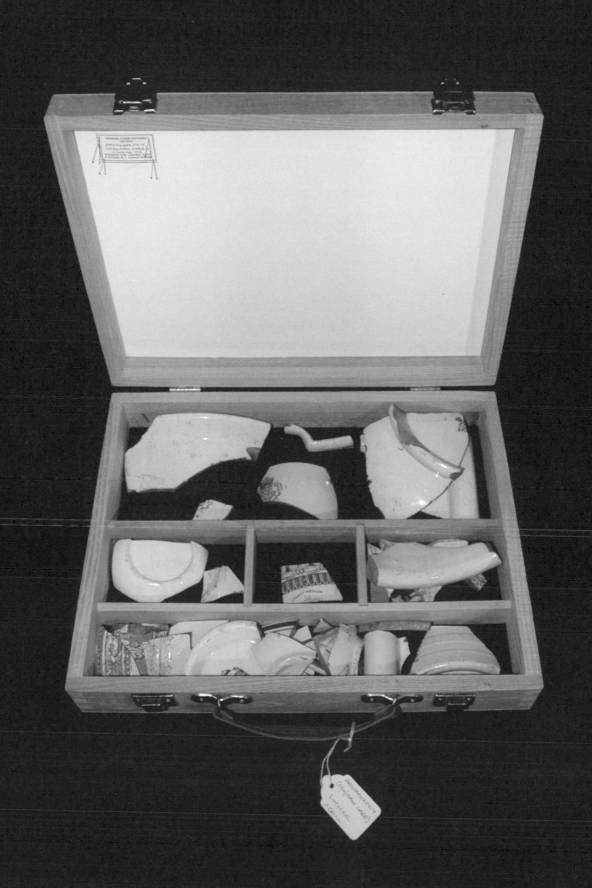

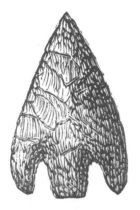

Archaeology as artistic practice
Archaeology is not a static practice:

> '…archaeologists do not happen upon or discover the past. Archaeology is a process in which archaeologists … take up and make something of what is left of the past.' (Shanks, 1992:50).

Like art it is therefore dynamic and ever-changing. According to this definition of archaeology it is an active process occurring in the present, where it is a cultural phenomenon in its own right, one that is interested in social practices and cultural phenomenon in the past, and what is left of them (Barrowclough, 2005).

If practitioners are to exploit the potential of the activity we call archaeology, they need to consider new approaches that look beyond their traditional links to anthropology or historical geography, to fields of practice whose concern has been with developing new ways of looking at the world. Contemporary art is one such field. As a social practice, contemporary art has the potential to offer new and useful insights into the past because it shares two key features with archaeology. Firstly, they are both modes of cultural production that work with material and intellectual resources to create contemporary meaning; and secondly, both are social practices, concerned with the negotiation of identities of people and things.

The performance of father and son in *Thesaurus Scienta Lancastriae* mimics the practice not only of archaeologists, but also of naturalists and geologists. To do so questions the special skill of the expert, subverting the authority of the professional. My first response to the challenge posed by *Thesaurus* would be to defend the privileged position of the archaeologist, to point out that what the Williamses have done bears only a superficial resemblance to the work of the practising archaeologist. While it appears scientific, in truth it is not. The skill of the professional lies in the ability to discriminate between objects on the basis of expert knowledge, rather than personal fancy. An archaeologist interested in the Bronze Age will have learnt the tell-tale signs that mark a flint arrowhead of that period from those of earlier periods, that separate Bronze Age swords from those of the Iron Age, Roman or Viking ages. In contrast, Jack's approach to collection seems indiscriminate.

Where archaeologists collect cultural artefacts Jack has felt no such constraint. His approach to collection has been all embracing, it has included elements of the natural world: plants, animals and minerals alongside the detritus of the modern world: metal, plastic and ceramic. An archaeologist might dismiss this approach, seeing in it naivety reflecting Jack's youth, but to do so I think would be to overlook a point of some significance. Unencumbered by the enculturation that all archaeologists are imbued with, Jack has approached collection with an innocence that few archaeologists can match. Through a child's eyes we are induced to reconsider the issue of what we collect and what we leave behind.

Natural collection

In 2001 I participated in fieldwork whose aim was to challenge archaeologists' preconceived ideas about what we collect and what we ignore. Under the direction of Cornelius Holtorf, the Monte da Igreja Project (see: http://members.chello.se/cornelius/Igreja/) was a biographical study of a Neolithic passage-grave. That is to say that evidence from all periods was considered equally interesting, and thus taken equally seriously. Studies that investigate 'life histories' (or 'biographies') of prehistoric sites have recently begun to attract interest within the archaeological community. The aim of such studies is to emphasise how prehistoric monuments impose an enduring presence on the landscape which invites successive generations of people to re-use and re-interpret the site (Holtorf, 2000–2001; Holtorf, 2001, Oliveira, 2000).

Through fieldwork we explored what it meant in practice to treat evidence from all periods equally. It meant that the Neolithic monument was no longer given primacy in the investigation, of equal interest were the foundations of a Roman farmhouse (excavated by Mary Chester-Kadwell and myself) a few metres to the north, and two geodetic markers – one derelict and dated to the mid-19th century, the other more recent and in use by, among others, modern-day archaeologists. All are presences in the landscape and therefore equally worthy of study. The study did not confine itself to the structures on the site surface, but also considered the artefacts that we found, both cultural and natural, over and in the site. These included used cans of Coca-Cola and shotgun cartridges, and an Arabic coin of the 11th century (identified by Trinidad Rico). The animal life that we encountered including ants, worms and, under a stone one day, a scorpion, was equally noteworthy. One of us (Ymke Mulder) was able to

THE UNDERWORLD
The Underworld was made in 1998 at the start of time, and will be gone at the end of time. It was also made in 2001, and then lots of magical talking animals were made.

identify the different species of plants growing on and in the site. All of these records went to form the document 'Monte da Igreja: a site inventory' (see: http://members. chello.se/cornelius/Igreja/inventory.htm), the data from which the life history of a small site in a rural part of Portugal could be reconstructed.

One of the most obviously novel aspects of this archaeological work was the recording of plants and animals alongside the more usual monuments and buildings. The distinction between Nature on the one hand, and Culture on the other seems unproblematic, but it has not always been so, and need not be so in the future. The Cartesian dichotomy arose through a particular set of social and historical circumstances that were peculiar to the development of Western thought following the Reformation (Scruton, 1995:27–37). The dualistic vision of the world first proposed by Descartes in the *Meditations of First Philosophy* (1641), which still guides much archaeological fieldwork, has been severely undermined by recent philosophy and by allied disciplines in the social sciences including anthropology (Strang, 1997). Archaeology, while aware of these developments (e.g. Thomas, 1996:11–30), has been slow to alter its practices. The reason may be that the breakdown of the Cartesian dualities represents an epistemological threat. The fear is that by collapsing the divide between Nature and Culture, objective standards of truth are lost, with relativism the consequence (Pearson & Shanks, 2001:98). One could argue, however, that the greater threat comes from not adjusting the archaeological mindset to take account of the particular historical circumstances that gave rise to the divide. Recent ethnography has demonstrated how non-Western (non-modern) peoples emphasise the relational character of their existence (Strathern, 1988) rather than the dualities of Cartesian philosophy. It was with this in mind that Holtorf established the Monte da Igreja Project and it is with this thought that I return to *Thesaurus Scienta Lancastriae*.

Park as artefact

Innocent of the philosophical debate, and left to his own devices, Jack has been able to compose his collection in a manner that seemed meaningful to him. His aim was to construct a collection of objects found in the park, and so he might think, 'why should I not collect both insects and plants, and pots and fireworks?' In so doing he has escaped the Cartesian dichotomy and taken an approach more familiar to people living outside the Western tradition. There a very different appreciation of the relationship between Nature and Culture is apparent. In Australia aboriginal belief does not make the distinction, neither does the Hindu religion in India, nor the beliefs of the San people of southern Africa. One of the problems that faces me as an archaeologist studying the prehistoric peoples of Britain is the nagging doubt that neither did they. The relationship between people and the world around them therefore becomes a central concern of a new investigation, what we might call a Life History Archaeology of Natural Places.

For an archaeologist of the natural place the park becomes the artefact, and all that is in it is deemed worthy of study. Freed from the separation of Natural and Cultural we can appreciate the relational character between people and the park around them. Here is an example inspired by the *Thesaurus*. The trees that line the avenues of the park include species from the four corners of the world, representing

the native species of the former British Empire. The original planting of the park was a celebration of British Empire, and says much about the attitudes and values of the upper middle class families of Victorian Lancaster. In the formal planting and arrangement of vistas we see a municipal version of the aristocratic country house gardens and parks, inspired by the aesthetic values of the aristocracy but paid for out of the profits of the expanding industrial capacity of northern England and presented to the community through the Christian values underlying public spirited charitable works.

As the years have passed since their first planting, so the trees have grown and matured, and with it the park and its buildings. The trees bear the scars of the on-going relationship between them and the people of Lancaster. From the simple heart inscribed by two lovers in a trunk to the more cryptic meaning of four coins deposited within a knarled bough we can see that the relationship between people and the parkland around them is on-going and often complex in its symbolism.

Conclusion

Thesaurus Scienta Lancastriae has appealed to the archaeologist within me because it has resonated with so many strands of thought. As is often the case it is only through the eyes of a child that one is able to distil so many of these into a meaningful idea. Fundamentally, the divide between Nature and Culture has to be viewed as a concept peculiar to Europe from the 17th century onwards. It cannot, therefore, be helpful to project this notion on to the lives of prehistoric peoples. What is required to understand past lives is an alternative approach, and what this might be is suggested by the *Thesaurus*. That is, to move away from a study of cultural artefacts *per se* towards a study that associates those objects with the 'natural' world, and which gives primacy to the *relationship* between people, and the world around them. The possibility of making this work has been ably demonstrated by the Williamses and the challenge of making this idea a reality for prehistoric Lancashire is the aim of my current research.

References

Barrowclough, D. A. 2005. 'Dancing in time: activating the prehistoric landscape of Lancashire' *Archaeological Review from Cambridge* 20.1.

Barrowclough, D. A. in prep. *A Modern Ethnography of an Ancient Past: interpreting the archaeology of later prehistoric North-West England*. Volumes 2, 3 & 4 Appendices Catalogues of Stone, Metal and Ceramic artefacts. PhD University of Cambridge.

Daniel, G. E. 1975. *150 Years of Archaeology*. Duckworth: London.

Descartes, R. 1998 [1641]. *Meditations of First Philosophy.* (Trans. Desmond M. Clarke) Penguin Books: London.

Holtorf, C. 2000-2001. *Monumental Past. The Life-histories of Megalithic Monuments in Mecklenburg-Vorpommern (Germany)*. Electronic monograph. University of Toronto: Centre for Instructional Technology Development.

Holtorf, C. 2001. 'The life-history of a monument: a new field of research' In *Monte de Igreja, Many Sites Many Sights: Thinking the Past* http://members.chello.se/cornelius/Igreja/introduction.htm.

Oliveira, Catarina 2000. 'Lugares de memória. Testemunhos megalitíticos e leituras do passado em Montemor-o-Novo'. *Historia* (February), 56-67.

Pearson, M. and Shanks, M. 2001. *Theatre/Archaeology*. Routledge: London, 98.

Schnapp, A. 1996. *The Discovery of the Past: the origins of archaeology*. British Museum Press: London.

Scruton, R. 1995. *A Short History of Modern Philosophy, from Descartes to Wittgenstein* (Second edition). Routledge: London, 27-37.

Strang, V. 1997. *Uncommon Ground: Cultural Landscapes and Environmental Values*. Oxford: Berg.

Strathern, M. 1988. *The Gender of the Gift*. University of California Press: Berkley.

Thomas, J. 1996. *Time, Culture and Identity, an interpretive archaeology*. Routledge: London, 11-30.

IMAGE REFERENCE

PAGE	PICTURE	COPYRIGHT
31	*Robert & Jack in white coats*	Gina Aylward
36	*Robert & Jack lying down near pond*	Gina Aylward
39	*Thesaurus Scienta Lancastriae*	Michael Coombs
40	*Robert & Jack looking at an insect*	Gina Aylward
74	*Sir Richard Owen & granddaughter*	Lancaster City Museum
79	*Edward Frankland*	Lancaster City Museum
80	*Thesaurus Scienta Lancastriae Interior*	Michael Coombs
83	*William Turner*	Lancaster City Museum
87	*Old postcard Greg Observatory*	Lancaster City Library
88	*The Cooke telescope*	Lancaster City Library
91	*Mr. Dowbiggin & small telescope*	Lancaster City Library
97	*The Palm House postcard*	Lancaster City Library
98-99	*Aerial view of Williamson Park*	Lancaster University Photographic Unit